THE INSTANT EGGHEAD

GUIDE TO THE MIND

EMILY

ANTHES

·

WITH A FOREWORD BY STEVE MIRSKY

THE

INSTANT

EGGHEAD

GUIDE

TO THE

MIND

ST. MARTIN'S GRIFFIN ✖ NEW YORK

ISBN-13: 978-0-312-38638-2

ACKNOWLEDGMENTS

A special thanks to Dr. Richard G. Pellegrino, M.D., for his expertise in fact-checking this book.

CONTENTS

Foreword   xiii

CHAPTER ONE
ALL IN YOUR HEAD: BRAIN STRUCTURE   1

Neurons   2

Neuron Communication   4

Glia   6

Spinal Cord   8

Brain Stem and Cerebellum   10

Thalamus, Hypothalamus, and Pituitary   12

Cerebrum   14

Cerebral Cortex   16

Hemispheres of the Brain   18

Whole Brain   20

Brain Imaging   22

CHAPTER TWO
MINDING YOUR BUSINESS:
BASIC FUNCTIONS   25

Body Regulation   26

Circadian Rhythms   28

Sleep   30

Dreaming   32

Eating   34

Sex   36

Movement   38

Mirror Neurons   40

Touch   42

Sensing Pain   44

Sensing Temperature   46

Smell   48

Pheromones   50

Taste   52

Hearing   54

Vision   56

Synesthesia   58

Stress   60

CHAPTER THREE

GOING MENTAL: HIGHER FUNCTIONS   63

Learning and Memory   64

Types of Memory   66

False Memories   68

Emotion   70

Fear   72

Happiness   74

Humor and Laughter   76

Love and Attachment   78

Creativity   80

Language   82

Language Acquisition   84

Decision-making   86

Economic Decision-making   88

Moral Decision-making   90

CHAPTER FOUR
YOU BREAK IT, YOU BUY IT: PROBLEMS
IN THE BRAIN   93

Autism,   94

Epilepsy   96

Attention-Deficit/Hyperactivity Disorder   98

Dyslexia   100

Depression and Bipolar Disorder   102

Anxiety Disorders   104

Phobias   106

Post-traumatic Stress Disorder   108

Amnesia   110

Schizophrenia   112

Addiction   114

Pathological Aggression   116

Sleep Disorders   118

Prion Diseases   120

Multiple Sclerosis   122

Parkinson's Disease   124

Alzheimer's Disease   126

Stroke   128

Headaches and Migraines   130

Head Injuries   132

Coma   134

CHAPTER FIVE
IT'S (USUALLY) NOT BRAIN SURGERY:
MENDING THE MIND   137

Psychotherapy   138

Psychopharmacology   140

Brain Surgery   142

Electroconvulsive Therapy   144

Deep Brain Simulation   146

Transcranial Magnetic Stimulation   148

Stem Cells   150

Gene Therapy   152

Robotic Limbs   154

CHAPTER SIX
THE POWER OF NURTURE: CHANGING BRAINS   157

Neuroplasticity   158

Neurogenesis   160

Exercise   162

Diet   164

Stimulants   166

Depressants   168

Other Drugs   170

Neurotoxins   172

Child Abuse and Neglect   174

Combat   176

Video Games   178

Music   180

Meditation   182

CHAPTER SEVEN

A MIND OF ONE'S OWN: INDIVIDUAL BRAINS   185

Genes and the Brain   186

Intelligence   188

Personality   190

Gender and the Brain   192

Sexual Orientation   194

Sex Hormones   196

Fetal Brain   198

Infant Brain   200

Teenage Brain   202

Parental Brain   204

Aging Brain   206

Evolution   208

Animal Minds   210

Consciousness   212

Appendix A: Commonly Prescribed Psychiatric
            Medications   215

Appendix B: Mental Health Resources   217

Appendix C: Recommended Reading   219

FOREWORD

When you think of *Scientific American,* what probably comes to mind is a long, impenetrable magazine article given as a reading assignment by a high school science teacher. The book you're holding now should change that perception. *Scientific American* has expanded, which is to say, contracted. We also do shorter, more penetrable writing to bring you up to speed so that someday, if you so desire, you will be *able* to tackle the long articles and find them penetrable, too.

A major new *Scientific American* offering in this vein is our podcast, "60-Second Science." This daily report represents a truly state-of-the-art technological breakthrough in the gathering and reporting of science news. That is to say, it's all done using a few hundred dollars' worth of equipment that fits in a carry-on bag.

But that's the magic of media today—thanks to the Internet, I can manage a worldwide science news operation from a bare-bones studio in my home. A stable of regular freelancers send ideas, mostly based on research in current journals. They then write and record the stories that I approve. I edit their voice files into the finished podcasts available at our Web site—freshly minted science piped directly into your ears.

This fast and fun science news effort has been a big hit. Now we offer the Instant Egghead guides to make foundational scientific information fast and fun as well.

The Instant Egghead Guide to the Mind is the ultimate user manual for that three-pound network of neurons (and other important stuff) that's busy keeping all your vital body functions in action, as well as making it possible for you to read and understand these words. It explains how your particular brain manages to do all those amazing things, while at the same time keeping important information—such as your spouse's birthday, the winner of the 1927 World Series, and your eighth-grade gym locker combination—available the moment you need it. (Actually, if you can't quite recall these things, you can find out why in here, too.)

So give your brain a treat—read *The Instant Egghead Guide to the Mind* and let yours get to know itself.

—Steve Mirsky

ALL IN YOUR HEAD: BRAIN STRUCTURE

NEURONS

Neurons are the building blocks of the brain. Your every perception, thought, and behavior can be traced back to them. At their most basic level, neurons are just cells. They have "cell bodies," which contain a single nucleus and other typical cellular "guts." But unlike other cells in the body, neurons are built for communication. Projecting from one end of the cell is a long tentacle called an axon. Its main purpose is to transmit messages to other neurons, like a telephone cable. The other end of a neuron has branches like a tree; these are dendrites, and they receive messages from the axons of other neurons.

All neurons share these basic properties, but otherwise there's a wide variety of cells in our bodies. Sensory neurons detect pressure and temperature, and send signals about those sensations, as well as pain, back to the brain—the electrochemical equivalent of "Yes, that mug is still way too hot to pick up."

Motor neurons, on the other hand, bring the brain's messages to the muscles, instructing those in your arm, for instance, to put the mug down right away.

Finally, there are interneurons, which mediate between sensory and motor neurons, as well as communicating among themselves. All of the neurons in the brain are, in fact, interneurons. Even these vary widely in form. Some have one or two dendrites; others have a thousand.

ON THE FRONTIER

Brain researchers are currently studying how neurons grow. Studies have now revealed the substances that help axons lengthen and the way new dendrites sprout from brain cells. Eventually, this research will not only show how neurons develop in the fetal and infant brain, but also help treat disorders like Alzheimer's, in which neurons become diseased or damaged. In fact, researchers have already designed synthetic substances that encourage new brain cells to form.

COCKTAIL PARTY TIDBITS

- ❖ No one knows for sure how many neurons there are in a human brain. (Would *you* want to count them?) But many experts put the number at about one hundred billion. If you spread out the cell membranes of all those neurons, you'd cover four soccer fields. (Yes, that would make it hard to play.)
- ❖ The longest axons in the human body—the ones that stretch from the spinal cord all the way to our toes—can be three or four feet long. In giraffes, those same axons are about fifteen feet long.
- ❖ Neurons are inconceivably tiny. On average, tens of thousands of them can fit on the head of a pin.

NEURON COMMUNICATION

Looking at individual neurons is almost pointless. What matters isn't what any one neuron does alone as much as what a bunch of them do together. Together, neurons form large networks, with messages constantly passing from cell to cell. The dendrites on a neuron receive a flurry of positive and negative signals, and the sum of these signals determines the charge that zips down the length of the axon, like baseball fans doing the wave down a row of bleachers.

By themselves, axons are not good conductors of electricity—worse than the wires in your house. But many axons are swaddled in myelin, a fatty coating that insulates the axon and helps the charge move rapidly down its entire length. Once the charge arrives at the tip, the axon releases one of several chemicals, called neurotransmitters.

Neurotransmitters are released into a small gap, called a synapse, that separates one neuron from its neighbor. The axon is on one side of the synapse and a dendrite of the receiving neuron is on the other. The neurotransmitter binds to receptors on the next neuron's dendrite, and the cycle starts all over again.

Through this combination of electrical signaling (down the length of the axon) and chemical signaling (across the synapse), the neurons signal one another without actually touching.

ON THE FRONTIER

Now that you've wrapped your head around the basic concept of neuron signaling, imagine having an artificial neuron slipped in between two real ones. That's exactly what scientists are working to accomplish. They have created microchips sporting artificial synapses that release neurotransmitters when stimulated, just like neurons. Implanted in the brain, these artificial synapses could theoretically signal the dendrites of your real neurons, doing the work of home-grown brain cells. The possibilities are staggering.

COCKTAIL PARTY TIDBITS

❖ Some neurons are connected to only one other nerve cell. Others are connected to as many as a hundred thousand.

❖ Nerve signals in some organisms can travel more than two hundred miles per hour.

❖ The squid's famous (to neuroscientists, anyway) escape reflex is controlled by an axon with no myelin at all. But because the axon is so wide—about one millimeter in diameter, which is huge for an axon—the signal can travel fast enough to wiggle the squid out of harm's way in the blink of a beady black eye.

GLIA

Neurons, like all self-respecting celebrities, need an entourage. In the brain, glial cells are that entourage. Glial cells, or glia (from the Greek word for *glue*), are the underappreciated, well, glue of the brain. Discovered in the nineteenth century, glia are the support cells packed around the brain's neurons.

For many decades, glia were defined by what they weren't: neurons. Unlike neurons, glia do not generate electrical signals and therefore scientists thought they had little influence on brain signaling. But just because they don't send impulses doesn't mean they're trivial. If neurons are the brain's CEOs, glia are their personal assistants, bringing them nutrients and oxygen, protecting them from pathogens, and maintaining the proper environment—pretty much everything but picking up their dry cleaning. There are many different kinds of glia, each specialized for one or more of these functions.

New research suggests that, like many personal assistants, glia haven't been getting enough credit. It turns out that glia don't just take orders from neurons but actually communicate with them. Studies have shown that glia can influence the connections that neurons make and help determine whether those connections get stronger. These findings mean that the long-overlooked glia might play a crucial role in the fundamental workings of learning and memory. How's that for a comeback story?

Stem cells are important because they can grow into many different kinds of specialized adult cells, making them a promising option for repairing damage to the nervous system. Now glia seem capable of the same effect.

In a study, scientists extracted glia from human brains and bathed them in a cocktail of proteins. When they transplanted the cells into the brains of mice, they turned into healthy adult neurons.

COCKTAIL PARTY TIDBITS

❖ Studies of the brains of other animals have revealed that species higher up on the evolutionary ladder have higher concentrations of glia.

❖ After Einstein's death, scientists discovered that his brain had a normal number of neurons. But Al had an unusually high concentration of glia. Perhaps he was more highly evolved than the rest of us?

SPINAL CORD

The spinal cord is the one and only highway to the brain—signals traveling to or from the brain have to traverse it. Motor neurons carry the brain's instructions through the spinal cord to the muscles, and sensory neurons ferry information back along the spinal cord to the brain. The cord itself is made up of neurons, and is surrounded by vertebrae, hard bones that protect the delicate nerves.

The best illustration of the vital importance of the spinal cord lies in the devastating results of a spinal cord injury. If the spinal cord is the highway to the brain, a spinal cord injury is like the only bridge collapsing. Signals stop in both directions, causing paralysis—the brain can neither tell the limbs how to move nor get information from them about the sensations they are feeling.

The higher up on the spinal cord the injury is, the more severe the repercussions. Injuries at the level of the chest will generally paralyze the legs. Spinal cord injuries in the neck can cause paralysis of all four limbs (quadriplegia), and injuries at the very top of the neck can impair basic functions like breathing. Christopher Reeve had an injury at the highest point along his spinal cord and couldn't breathe on his own. At least at first. A real-life Superman to the last, Reeve gradually worked up his stamina until he could breathe without his ventilator for brief periods.

ON THE FRONTIER

It was long assumed that patients with spinal cord injuries couldn't make any meaningful recovery, and they were encouraged to merely come to terms with their new physical limitations. But scientists are now demonstrating that this fatalistic assumption is incorrect—many patients can regain or at least improve their mobility.

A demanding course of rehab, in which medical aides guide paralyzed patients through the motions of walking on a treadmill, has been shown in isolated circumstances to improve patients' walking abilities. Scientists believe that this rehab might strengthen any nerve or spinal cord connections that remain and rewire important neural circuits.

COCKTAIL PARTY TIDBITS

❖ Humans and giraffes have the same number of vertebrae in their necks: seven apiece. Each of those seven giraffe vertebrae can be almost a foot long.
❖ The human spinal cord is about seventeen to eighteen inches long—substantially shorter than the length of our spines.

BRAIN STEM AND CEREBELLUM

The brain stem is not the most glamorous of structures, but it is one of the most important. Attached to the top of the spinal cord, the brain stem controls a number of life's basic functions, functions not under our conscious control. It maintains heart rate and blood pressure, regulates breathing, and controls digestion.

The brain stem is also responsible for maintaining levels of alertness and consciousness. As part of this task, it helps filter the sensory information constantly bombarding us, determining where we should direct our attention and what we can safely ignore. The cranial nerves, which control basic movements of the head, face, and neck, are also part of the brain stem.

The cerebellum is located at the base of the brain, just behind the brain stem. It actually looks like a miniature version of the brain—wrinkles, hemispheres, the whole shebang. (In fact, *cerebellum* actually means "little brain.") But while the "big brain" is involved in a wide variety of functions, the cerebellum specializes in sensing the body, its movements, and its location in space.

The cerebellum is responsible for our sense of balance (or lack thereof), and is what keeps us standing (or not) after a few drinks. The cerebellum receives messages about what your body is doing and helps coordinate your movements appropriately in a continuous feedback loop. Don't leave home without it.

ON THE FRONTIER

Eating, drinking, and peeing (yes, peeing) are so essential that the brain stem has a special mechanism that allows the body to perform these actions even when in severe pain. According to a new study, the brain stem can suppress pain long enough for animals to complete behaviors crucial to their survival.

Researchers figured this out by putting rats in cages with floors that could get hot enough to cause pain. Usually the rats would pull their paws away from the heat, but while they were eating, the brain stem actively suppressed the pain, and the rodents left their paws on the hot floor long enough to satisfy their hunger.

COCKTAIL PARTY TIDBITS

- ❖ Even though the higher brain centers process emotion, it's the brain stem, with its cranial nerves, that allows you to make the complex series of movements necessary for a smile.
- ❖ The cerebellum is 10 percent of the brain's total volume but contains more than 50 percent of its neurons.
- ❖ Alcohol abuse is a common cause of damage to the cerebellum.

THALAMUS, HYPOTHALAMUS, AND PITUITARY

The thalamus is the brain's switchboard operator, routing messages to and from the areas of higher functioning in the brain. (In all likelihood, you've never dealt with a switchboard operator, but the analogy stands. Check Wikipedia, if you have to.) There are two thalamuses (technically, thalami) in the brain, one in each hemisphere.

Beneath the thalamus (I just can't bring myself to say *thalami*) is a small structure called the hypothalamus (*hypo-* meaning "under"). It helps regulate some of the body's most basic—and fun—activities: eating, drinking, and sexual behavior (thank you, hypothalamus). It also plays a vital role in the body's stress response, particularly by activating the pituitary gland.

The pituitary gland, located at the base of the brain, is a pea-size gland in charge of the body's endocrine, or hormone, system. The hypothalamus tells the pituitary gland when to produce hormones; those released by the pituitary control growth, sexual function, metabolism, and more. (Awkward teenage years? Blame your endocrine system, which floods your body with hormones during adolescence that cause growth spurts, sexual maturation, voice changes, acne, and a host of other indignities.)

Because of its role in regulating appetite and eating, the hypothalamus has been of great interest to researchers studying obesity. The structure modulates how much we eat and how fast we burn it by monitoring the levels of various nutrients in the blood.

A relatively recent finding indicates that the hypothalamus also keeps track of the body's fatty acid levels and adjusts appetite accordingly. When researchers lowered the level of these acids in mice, the rodents ate more and became obese.

This finding suggests that increasing levels of these acids in the body might reduce appetite. Scientists hope to harness this phenomenon to help curb human obesity.

COCKTAIL PARTY TIDBITS

❖ Deficiencies in vitamin B can damage the thalamus, causing a disorder known as Korsakoff's syndrome. A hallmark of Korsakoff's syndrome is that patients typically invent new memories to fill in the gaps left by their amnesia.

❖ There are differences between the male and female hypothalamus—which has been implicated in both sexual orientation and gender identity. Research has shown that giving female sex hormones to a young male rat can "feminize" his hypothalamus and vice versa.

CEREBRUM

The cerebrum, ladies and gentlemen, is the Big Kahuna, the biggest of the brain's sections and home to most of the functions (and dysfunctions) that make us human.

Love, language, your inexplicable fondness for Celine Dion—that's all the cerebrum. This gray walnut is what people think of when they think of "the brain." The cerebrum consists of both white and gray matter. White matter is the name given to regions that consist primarily of axons covered in myelin (the fatty, insulating substance that keeps signals traveling smoothly, remember?). The myelin makes that brain tissue white. Many of the deeper layers of the cerebrum consist of this white matter. Gray matter, on the other hand, is composed mostly of neuron cell bodies, which aren't covered in myelin.

The cerebrum contains a number of important structures that we'll discuss. A quick rundown of the most notable: the amygdala, an almond-shaped structure involved in emotional processing, particularly fear; the hippocampus, which is crucial to long-term memory; the basal ganglia, a group of nuclei that help monitor and control movements; the relatively thin outer layer of the cerebrum, the cerebral cortex, which has so much going on that it merits its own spread. Turn the page for more.

A study of white matter has revealed that it might be far busier than scientists initially imagined. The prevailing wisdom has been that gray matter does the brain's heavy lifting and is the main site of information processing. White matter, on the other hand, has been viewed as a mere conduit, carrying the gray matter's signals from point A to point B.

But new research shows that the axons present in the brain's white matter aren't just passing along messages from gray matter—they're also processing information themselves (good for them!). It's just one more example of how much we don't know about the most basic workings of the brain.

COCKTAIL PARTY TIDBITS

- ❖ The cerebrum makes up 85 percent of the weight of the human brain.
- ❖ *Hippocampus* comes from the Greek words for "seahorse," which seems bizarre until you see that it is actually shaped like a seahorse. So the name actually makes a lot of sense. Learn your Greek and Latin roots, kids!
- ❖ People who are tone deaf seem to be missing some white matter, according to one study. No word yet on whether karaoke bar patronage helps spurs white matter growth.

CEREBRAL CORTEX

THE BASICS

The cerebral cortex is the thin top layer of the cerebrum. Only two millimeters thick, this sheet of gray matter is intricately wrinkled and folded, a living piece of origami. The folds drastically increase the surface area. (Imagine fitting a large piece of paper in a small box by crumpling it.) In fact, if you unfolded the whole cortex, it would be about two and a half square feet.

The cortex is commonly divided into four regions, called lobes. The frontal lobe (located in the, um, front of the cortex) is home to most of the higher mental functions we associate with being human, including personality and decision-making. It also contains the motor cortex, which controls voluntary movements. Behind the frontal lobe is the parietal lobe, which contains the sensory cortex. The sensory cortex gets messages from the body's sensory organs and neurons.

Just under the parietal lobe, around the area of your ear, is the temporal lobe. As you might guess from its location, that lobe is involved in processing speech. Finally, there's the occipital lobe, all the way at the back of the brain. The occipital lobe's main task is to process visual information. So, in a way, we all have eyes in the backs of our heads.

It takes time for the twists and turns of the cortex to develop, and that process occurs over the course of fetal and infant development. Researchers now have a model for when and how this process unfolds (pun, well, intended). When a fetus is thirty-two weeks old, its brain is practically smooth. Between weeks thirty-three and thirty-eight, many of the folds appear, and by thirty-eight weeks after conception its brain will look almost as wrinkled as an adult's.

This timeline will help scientists and doctors closely monitor brain development and potentially predict abnormalities at an extremely early age.

COCKTAIL PARTY TIDBITS

❖ The folds of the cortex aren't simply called folds. The peaks are gyri (a single peak is a gyrus) and the valleys are called sulci (singular, sulcus). Mind your gyri and sulci.

❖ The human cortex is elaborately folded, but many animals have cortices that are practically smooth. Our wrinkled cortex might be what gives us *Homo sapiens* such mental prowess. Finally—wrinkles we can celebrate.

HEMISPHERES OF THE BRAIN

The cerebrum is divided into the right hemisphere and left hemisphere. The right hemisphere controls the left half of the body, while the left hemisphere controls the right side of the body. But the hemispheres need to coordinate their actions and share information. A thick bundle of fibers, the corpus callosum, joins the two hemispheres, allowing them to communicate.

There's a lot of pseudoscience swirling around the idea of hemispheric dominance. It's true that one half of a person's brain will be slightly more dominant than the other. And there are some differences between the hemispheres, particularly in language processing.

But the idea that the right half of the brain is the creative side and the left half is the logical side, for instance, is greatly oversimplified. Most brain functions take place on both sides of the brain, and studies of patients with brain injuries in one hemisphere have shown that the other can compensate. Those online tests that tell you whether you're "right-brained" or "left-brained"? All they prove is that you need to find a better way to procrastinate. Try YouTube.

A recent study of capuchin monkeys, those adorable South American primates, turned up some interesting factors that influence the size of the corpus callosum. Sex, for one. The researchers stuck the monkeys in MRI machines—they don't have those in the jungle—and found that the corpus callosum was smaller in males than females. Handedness played a role, too. The bundle of fibers was smaller in right-handed monkeys than in southpaws. A larger corpus callosum may mean that the brain's hemispheres work together more closely.

COCKTAIL PARTY TIDBITS

❖ Some of the most interesting neuroscience studies involve split-brain patients, who have their corpus callosums cut. This leaves the hemispheres unable to talk to each other and creates strange behavioral symptoms in which patients seem unaware of what half of their bodies are experiencing.

❖ Even more drastic is a procedure called a hemispherectomy, in which doctors remove an entire hemisphere of the brain. This procedure, performed almost exclusively in children, is used to treat severe epilepsy or seizure disorders. If the patient is very young, the hemisphere that remains can take over most of the brain's functions, and the patient can lead a relatively normal life. Pull out that piece of trivia next time someone calls you a half-wit.

WHOLE BRAIN

Imagine an egg. Visualize its hard shell, the dense yolk suspended in a the liquid egg white. This is your brain. (Don't worry—there's no impending analogy involving a frying pan.) Like an egg, your brain is protected by a hard shell (your skull) and a thin layer of clear liquid (cerebrospinal fluid) which keeps it from banging around inside the skull. The brain is also cushioned by a few thin membranes that help absorb the shock of sudden movements. (Mother Nature's version of Styrofoam peanuts.)

The human brain, all told, is about three pounds in weight and the size of a grapefruit. Despite what you may have seen in museum exhibits, it is not actually hard and gray. The living brain is red and soft; the chemicals involved in preserving a brain dull its appearance. Though it represents only a small percentage of the body's weight, the brain receives up to a fifth of the blood pumped by the heart. The brain is an organ, just like the lungs or liver, and requires tremendous amounts of nutrients, oxygen, and fuel—in the form of the sugar glucose. Arteries weave in and around the brain, delivering these essentials. Glial cells turn the glucose into a fuel that neurons can run on.

Cerebrospinal fluid is already widely used to diagnose infections. (Doctors take a sample of the fluid with a spinal tap to analyze it for the biological indicators of an infection.) But in the future, we might examine it to diagnose psychosis. Research has shown that psychotic patients have abnormal levels of several compounds in their cerebrospinal fluid. These abnormalities were present only in people with diagnosed psychosis—not in healthy volunteers or patients with other mental illnesses. Scientists still aren't sure what causes these anomalies in the fluid (and whether they cause psychosis or are a symptom of it), but the research continues.

COCKTAIL PARTY TIDBITS

- The heaviest human brain—as recorded by the *Guinness Book of World Records*—tipped the scale at 5 pounds, 1.1 ounces.
- The skull figured prominently in the pseudoscience of phrenology. Practitioners claimed to be able to discern personality traits by feeling the bumps on people's heads.
- The brain uses one-tenth of a calorie a minute. Thinking hard can increase the burn, although *The Thought Diet* probably won't be hitting bookshelves any time soon.

BRAIN IMAGING

It's hard to know exactly what's going on in someone's head at any given moment (especially if that someone is your significant other). Fortunately, we have a variety of techniques that allow researchers to peek inside our skulls. Unless you're writing a textbook, all you really need to know are the acronyms:

- ❖ **CT:** A CT machine uses an X-ray that revolves around the head, creating images of various slices of the brain. No actual slicing required.

- ❖ **EEG:** EEGs measure electrical activity in the brain over time, most useful for measuring changes in consciousness like sleep or coma.

- ❖ **MRI:** MRI machines create a powerful magnetic field that aligns the hydrogen atoms in the brain. As they return to position, they emit signals that can be imaged. Regular MRIs provide information about shape and structure of the soft tissues that CT scans miss, and functional MRIs can track activity in different brain regions by measuring blood flow.

- ❖ **PET:** Doctors inject a radioactive substance that travels to the brain and collects in the areas working the hardest. The radioactivity helps map brain activity from moment to moment.

Rapidly advancing technology has led to the development of several new techniques for imaging the brain. The latest tools allow scientists to watch the workings of *single* neurons. Researchers can now capture images of the firing of individual neurons and communication in neural circuits, rather than having to rely on images of brain areas as a whole.

These imaging techniques typically work by detecting the flow of calcium into a neuron—one of the steps that precedes neuron firing. Eventually, they might reveal the cellular basis of everyday mental functions and shed light on what goes wrong in damaged neurons.

COCKTAIL PARTY TIDBITS

- ❖ MRI machines aren't just for brains. The first MRI scan of a human, completed in 1977, was a cross-section of the chest.

- ❖ The magnets in most MRI machines are more than ten thousand times as strong as Earth's magnetic field. That's why they tell patients to remove any jewelry or piercings before a scan. Especially piercings.

- ❖ Forget the Nintendo Wii. Several companies are creating video games with controllers inspired by EEGs. Helmets detect brain activity, allowing players to control games by thinking of simple actions.

TWO

MINDING YOUR BUSINESS: BASIC FUNCTIONS

BODY REGULATION

Before emotions, creativity, or insight, the brain tends to the more mundane needs of the body, largely through the hypothalamus, which works with the brainstem to regulate the body's autonomic nervous system. This is stuff that operates below the level of conscious thought: heart rate, breathing, digestion, and so on.

The hormones produced by the hypothalamus and the nearby pituitary gland trigger endocrine glands all over the body to produce hormones of their own. These hormones influence metabolism, muscle growth, and lots more. And the hypothalamus regulates some of our most vital behaviors, such as eating.

But the brain can't just give orders blindly—it needs feedback from the body. This feedback comes in several ways. The first is through the blood supply. The brain can tell a lot about what's going on in the body by monitoring the oxygen, hormones, and nutrients in the blood it receives. Various nerves in the body also send information to the brain. One of the most important of these is the vagus nerve, which transmits information about the stomach and the heart. What happens in the vagus nerve doesn't stay in the vagus nerve.

Researchers are finding evidence that the brain influences the immune system, an idea scientists used to scoff at. But according to some studies, stress hormones can suppress the immune response, making the body more susceptible to disease.

And that's just the beginning of the brain–immune system link. The brain appears to receive messages from damaged or infected tissues and then influence the body's response to that damage. And immune-system molecules that reside primarily in the brain seem to be involved in all sorts of neurological brain diseases.

COCKTAIL PARTY TIDBITS

- Stimulating the vagus nerve might help disorders ranging from epilepsy to depression. How sending electrical signals to the vagus (and then on to the brain) helps someone's mood is unclear, but it does help. At least, the FDA seems to think so: The agency approved the treatment for depression in 2005.
- Disabling the vagus nerve causes a loss of the gag reflex. Useful for watching reality TV.
- Fevers are controlled by the hypothalamus—another example of the feedback loop between the brain and immune system.

CIRCADIAN RHYTHMS

Forget what your band teacher told you. You've got rhythm. Lots of rhythms, actually. The brain has to make sure that all the parts of the body are doing their jobs at the right times, according to a daily cycle. The master coordinator of these rhythms lies in a small group of neurons in the hypothalamus called the suprachiasmatic nuclei (or SCN).

The SCN ticks off a roughly twenty-four-hour cycle using messages from the eyes to synchronize itself with light and dark. The SCN then acts as the conductor of the body's orchestra of functions, coordinating them and keeping them in time.

It's not just eating and sleeping patterns that follow circadian rhythms. Your body temperature has a circadian rhythm, as does your blood pressure, respiration, and pain sensitivity. Even your mental functions don't remain constant over the course of the day—your alertness peaks in the late morning, your mood at midday, your reaction time in the late afternoon.

Disturbing the body's natural circadian rhythms creates physiological havoc (like jet lag). Scientists have bred animals that have no circadian rhythms, and their dysfunctions range from metabolic disorders to infertility.

Guess it's better to live life by the clock.

Our clocks aren't all keeping the same time. Some of us are owls, staying up late and sleeping in. Others are larks, asleep before *The Daily Show*, awake before dawn. Your preferences might be inscribed in your DNA.

The difference has been tied to several slight variations in the genes expressed in the SCN. Scientists even found an entire family who tuck themselves in by seven P.M. each night and arise long before dawn. Turns out their schedules could be tied to a specific mutation. Those with the early bird gene get the worm.

COCKTAIL PARTY TIDBITS

- ❖ The human circadian rhythm is slightly longer than twenty-four hours. When volunteers are sequestered in rooms without natural light, their bodies shift to a day lasting about twenty-five hours. Light helps keep our clocks pegged to Earth's day.
- ❖ During total eclipses of the sun, cows and other animals will just lie down and go to sleep, even if it's the middle of the day.
- ❖ In 1962, French geologist Michael Siffre spent 205 days living alone in Midnight Cave in Texas to study what would happen to his circadian rhythms when he was utterly deprived of daylight. The experience wreaked havoc on his sleep patterns; it took years for him to recover.

SLEEP

In 1964, seventeen-year-old Randy Gardner embarked upon what might be the most badass science project ever: He stayed awake for eleven days straight, the longest period of documented human sleep deprivation.

When we're not setting world sleeplessness records, we humans spend about one-third of our lives asleep. Throughout the night, we cycle through five different sleep stages, each of which has its own distinct patterns of brain activity. As we progress from stage one to stage four, our sleep gets deeper. Stage four is followed by rapid eye movement (or REM) sleep, usually accompanied by dreaming.

Even slight sleep deprivation can take a serious toll. Missing out on sleep affects concentration and learning and can impair functioning as much as being drunk. Sleep deprivation can also compromise the body's immune response. And we'd die faster without sleep than without food.

But researchers still can't tell us why the need for sleep evolved. One theory suggests that sleep helps us learn and form memories, another proposes that it might repair cell damage. In all likelihood, sleep serves more than one purpose. Given the immense risks that accompany it—sleeping would have made our ancestors on the savannah less likely to find dinner and more likely to become it—sleep must be working some serious miracles.

ON THE FRONTIER

Scientists are now exploring the idea that sleep deprivation could be implicated in our great nation's great obesity epidemic. Sleep deprivation can affect the levels of certain hormones that regulate appetite and eating, which may cause those of us who don't sleep enough to pack on the pounds.

According to one study, people who don't get enough sleep are nearly twice as likely to be obese. Researchers say it's no coincidence that, as a society, we're sleeping less and eating more. When it comes to weight, the old adage might be true: You snooze, you lose.

COCKTAIL PARTY TIDBITS

- We first start yawning in utero. Bad news, fetuses: Life only gets more exhausting.
- The purpose of yawning remains a mystery, but one recent study suggests that it cools the brain, which could improve alertness and concentration. Others say that yawning serves more of a social purpose, signaling that it's time for everyone to relax.
- Dolphins sleep in one hemisphere of their brain at a time, allowing them to recharge without literally laying their heads down to rest.

DREAMING

To sleep, perchance to dream? Perhaps if you're in REM sleep, the sleep stage in which most dreams occur. Rapid eye movement sleep is characterized by (surprise!) the rapid movement of our eyes. Brain activity during REM sleep looks a lot like our brain activity when we're wide awake. Though our eyes are fluttering like mad, the rest of the body is paralyzed.

Research suggests that this sleep paralysis helps protect us from becoming physically involved in our dreams. (People who suffer from a disorder that keeps their bodies from becoming paralyzed during REM have been known to act out their behavior in their dreams, sometimes injuring themselves in the process.) REM periods start out relatively short but get progressively longer throughout the night.

No one knows exactly where in the brain dreams originate. Some have suggested that dreams are merely our brains' interpretations of random neural firing, while others believe that dreams are a byproduct of the memory processing that happens during sleep—if, I suppose, you have memories about showing up at work naked.

What dreams mean—indeed, whether they mean anything at all—is a long-standing scientific question. Experiments have shown that dream content can be influenced by external events, such as smells presented just before sleep.

One sleep researcher even showed that people who spent several hours during the day playing Tetris reported seeing Tetris-like shapes in their sleep.

ON THE FRONTIER

Having a baby can change everything. Including, it seems, your dreams. A study of new mothers reveals that they have intense dreams about their babies and that about three-quarters of these dreams are negative.

The dreams often feature anxiety-provoking scenarios in which their newborns are in danger. The women reported that they were anxious about their babies when they awoke and felt compelled to go check on them. Dads can have the bad dreams, too.

COCKTAIL PARTY TIDBITS

❖ In a normal night, we spend about a quarter of our time in REM sleep. Assuming you get eight hours a night, that means two hours of dreaming. Over the course of a year, that's a month of dreaming. (And that doesn't even include daydreams.)

❖ Being awakened during REM sleep increases your odds of remembering your dreams. If you want to remember more dreams, set an alarm to wake up an hour or two early, jot down any dreams you remember, then go back to sleep.

EATING

Chew on this: Appetite and eating are regulated by a suite of chemicals like gherlin, commonly referred to as the hunger hormone. Released by the stomach, gherlin levels spike several times throughout the day, whenever your body is expecting a meal. The hormone travels to the brain, hitting the hypothalamus and triggering the feeling of hunger.

As you eat, your stomach stretches, sending your brain signals that you might want to put down those fries. The body also relies on an appetite-suppressing hormone called leptin, which is produced by body fat. The more fat you have, the more hormone in your system. In theory, people with a lot of body fat should feel fewer hunger pangs than lankier folks.

If you're thinking leptin sounds like a potential diet drug, you're not the only one. Since its discovery in 1994, scientists have investigated using supplements of the hormone to suppress appetite in overweight patients. But it turns out that leptin is only effective in those few people whose bodies can't make the hormone properly. Looks like the grapefruit diet will have a market for some time to come.

Eating causes a surge of dopamine, a neurotransmitter, in the brain, which feels good. Recent research shows that obese people have fewer dopamine receptors than they should. This means that obese people may feel more compelled to do things that increase dopamine levels. Like eat.

Drug addicts also have relatively few dopamine receptors. Neuroscientists are beginning to accept that overeating is a form of addiction, with food acting as some individuals' drug of choice. Seeing overeaters as addicts might lead to promising new interventions for weight control.

COCKTAIL PARTY TIDBITS

- Much of what we know about the effects of food deprivation comes from the Minnesota Starvation Experiment. The study recruited conscientious objectors to WWII; the men volunteered for months of near-starvation as an alternative to serving in the military. The study revealed the drastic psychological effects of famine conditions. The thirty-six participants experienced depression, food obsession, self-mutilation, and more.

- In "gourmand syndrome," a patient suddenly develops an obsession with eating gourmet food. This rare syndrome can occur after injury to the right frontal lobe.

SEX

I know—you turned to this page first. Welcome! But proceed with caution: Nothing will dampen your sex drive faster than reading about the biology behind it.

Touching the genitals creates sensations (usually pleasurable) that are carried to the spinal cord. These signals provoke certain automatic physical responses (erections, for instance) as well as responses in the brain, particularly in the sensory cortex and the limbic system. Activation of the limbic system, which includes the hypothalamus, makes sex pleasurable and triggers the body to release hormones that influence sexual behavior and functioning.

Some lucky scientists are studying what happens in the brain during orgasm. Orgasms seem to activate the brain's pleasure center, a group of cells called the nucleus accumbens, which is involved in feelings of reward and pleasure— as well as addiction.

Recent research has shown that, in women, certain parts of the brain essentially turn off during orgasm. In particular, activity slows in the amygdala, the part of the brain responsible for fear and anxiety. (The researchers also studied men, but couldn't obtain worthwhile data because their orgasms were too short.) Overall, brain activity during orgasm resembles the brain's response to drugs, researchers say.

Researchers have begun using thermography to measure sexual arousal. Thermography shows temperature changes in the genitals as people get excited; research shows that increasing temperature reliably corresponds to how turned on people say they are.

Scientists have already used the technique to demonstrate that the thermal changes of arousal are similar in men and women and that—contrary to the popular stereotype—things heat up just as quickly for women.

COCKTAIL PARTY TIDBITS

- ❖ Unsurprisingly, perhaps, researchers have discovered that the deactivation of the brain's fear circuits—which occurs in women having real orgasms—doesn't happen when women fake it. Maybe it's time to tell your girlfriend your "fantasy" about doing it in a brain scanner.
- ❖ Semen quality, in terms of number of sperm, is highest in the afternoon. Couples trying to conceive may benefit from some afternoon delight.

MOVEMENT

So you want to do a cartwheel. Congratulations. Now your brain has to plan and execute the movement. The intuitively named premotor cortex and motor cortex help initiate voluntary movement, sending signals to the brainstem and spinal cord where the motor neurons are located. These neurons fire, telling your muscle fibers to contract.

Contracting muscles help you tentatively reach down to the ground and heave your legs over your head, a clumsy attempt at a cartwheel. Maybe not a pretty one, but your body got the message. Now you just need practice.

Of course, that's not all there is to movement. The motor cortex doesn't work alone. Its activity is influenced by several other parts of the brain, including the cerebellum, which monitors body position and balance, and the basal ganglia, which helps the cortex plan movement. And reflexes are a different beast altogether; they don't even use the brain. They are automatic and fast—handy, for instance, when you put your hand down on a hot stove—because they are controlled by the spinal cord rather than the brain. No time is wasted on conscious thought. You just pull back, and your brain can focus instead on finding the proper expletive.

Though aging ravages certain parts of the brain, the cerebellum is largely spared. The cerebral cortex shows a whole host of changes in gene activity as we get on in years. But in the cerebellum, these changes are far less profound. Given that the cerebellum helps regulate breathing and heartbeat, that's probably a good thing.

COCKTAIL PARTY TIDBITS

❖ Infants exhibit a handful of reflexes that adults do not, including one that makes them curl their little hands around anything that touches their palms.

❖ Different parts of the body are controlled by their own regions in the motor cortex. About a quarter of the motor cortex is entirely devoted to movement of the hands.

❖ Some of the most complex movements you make involve your mouth and tongue. Get your mind out of the gutter—I'm referring to speech, which involves the coordinated control of about 100 muscles.

MIRROR NEURONS

In the early 1990s, a team of Italian neuroscientists was measuring the brain activity of macaque monkeys. The researchers attached electrodes to individual neurons in the monkeys' premotor areas, a region involved in planning movement. When the macaques performed an action, such as picking up a grape, neurons in the region would fire.

To their surprise, the researchers discovered that certain neurons in this region fired not only when the monkey picked up the grape himself, but also when he watched another monkey do so. (Monkey see, monkey do, huh?) The researchers named these cells "mirror neurons" because of their apparent activity in mirroring the actions of others. Mirror neurons also respond to sound—the same neuron will fire when a monkey cracks open a nut as when he *hears* a nut being cracked open.

Many scientists believe these neurons play an important role in imitative learning. The mirror system may also help us recognize the intentions behind others' actions. Individual mirror neurons have yet to be directly observed in humans, but functional MRIs have found certain areas of the brain that are active both when a person performs a task and when that person sees someone else do it. These areas have been dubbed the human mirror neuron system.

Scientists have suggested that mirror neurons could explain everything from empathy to the evolution of language. One of the most provocative recent studies found that children with autism, a developmental disorder, have reduced mirror activity when watching and imitating human facial expressions. This suggests that a malfunctioning mirror neuron system might give rise to the social impairments that characterize autism.

These notions remain controversial, but research continues at a furious pace.

❖ Some people even suggest that mirror neurons explain why we like pornography. A study showed that viewing aroused genitalia activated mirror systems in the premotor cortex of the human brain. In other words, looking at sex triggers activity in some of the same brain regions that having sex does. Voilà, *Debbie Does Dallas*.

❖ Moving on from sex to violence, another study found that certain mirror systems turn on in children watching violent TV shows. Some interpret this finding to mean that the kids will be more likely to watch then commit aggressive acts themselves. Consume with the requisite grains of salt.

TOUCH

The average adult has a whopping twenty square feet of skin on his or her body. A variety of touch receptors are embedded throughout. Each kind of receptor is specialized for detecting a different kind of sensation: pressure, vibration, and so on. When a receptor senses a stimulus, it sends signals through sensory nerves to the spinal cord, which transmits them to the thalamus (the switchboard, remember?) and then on to the somatosensory cortex.

The somatosensory cortex contains a topographical "map" of the body. That is, one region of the cortex responds to touch sensations from the back, another area responds to sensation from the hand, and so on. One of the interesting things about this map is that the size devoted to each body part varies, not according to how big that body part is, but how sensitive.

The fingers, for instance, have a lot of touch receptors. Thus, they have a huge portion of the cortex devoted to their sensations. The trunk of the body, on the other hand, is much less sensitive and gets only a small sliver of the cortex.

Research has revealed that people who read Braille—a skill that is performed with the index finger—have much larger cortical regions devoted to receiving sensations from their index fingers than the rest of us do.

Interest is growing in a field known as haptic technology, which involves endowing machines and robots with a sense of touch. Scientists have already made a number of break-throughs. One has designed a robotic fingertip capable of sensing information about the surfaces of objects. Another is developing a robotic rat, which has tactile sensors on its "whiskers." Such robots might be useful for performing all sorts of dangerous tasks, from combing through debris on search and rescue missions to studying the surfaces of other planets.

COCKTAIL PARTY TIDBITS

* Those twenty feet of skin? They weigh about six pounds, all told.
* A small percentage of children have what's known as tactile defensiveness, which makes them hypersensitive to touch. These kids might have trouble finding clothes that feel comfortable on their skin, shy away from physical contact, or react strongly to certain textures.

SENSING PAIN

Pain, by definition, is unpleasant. But it serves a noble purpose, teaching us to avoid things that are dangerous. The perception of pain begins with nociceptors, special receptors in the body that react to anything strong enough to potentially harm the body's tissues (a prick or a burn, for example).

When the nociceptors detect one of these troubling sensations, they send messages to the brain—via the spinal cord—which receives the signal as ouch-inducing. If necessary, the brain can also send messages back down the spinal cord to tell your muscles to get the hell away from the stimulus, already.

Nociceptors usually only respond when they are stimulated strongly, but damaged tissues release chemicals that make the nociceptors much more sensitive. That's why it hurts just to touch sunburned skin.

Fortunately, we're not helpless to deal with our pain. When the body is under stress, the brain releases endorphins, which are neurotransmitters that make us feel better. Chemically, endorphins are part of a group of compounds known as opioids.

Know any other opioids? Morphine's one. Oxycodone is another. These substances, you may be lucky enough to know from firsthand experience, also interact with opiate receptors in the brain and work quite nicely in treating pain.

Chronic pain is an enormous medical problem, one that costs the United States as much as a hundred billion dollars a year. And as many as half of those suffering from it get no relief from the treatments currently available.

But new neuroscience research is yielding promising insights into chronic pain. Studies show that the brain responds differently to long-term and short-term pain. This helps explain why so many treatments appropriate for acute pain fail to help patients with chronic discomfort. It may also help researchers develop new, more effective interventions for long-term pain.

COCKTAIL PARTY TIDBITS

- Hot peppers contain a substance known as capsaicin, which, it turns out, triggers nociceptors, which is why hot food hurts.
- Rubbing your toe after you stub it is actually a sound medical intervention: The sensation of rubbing sends signals to the brain that interfere with the pain signals.
- It sounds enticing, but an inability to experience pain—which can be caused by a variety of medical conditions—can actually be quite dangerous. People who have little or no ability to feel pain can die from injuries or infections they never even feel.

SENSING TEMPERATURE

In addition to nociceptors, the body contains receptors that detect warmth and cold. The receptors work through a sophisticated cellular mechanism, but I'll just outline the basics: Cells in the skin contain specialized channels that open when they sense a temperature change.

These channels come in an array of types—some respond to heat and others to cold. In either case, the opening of the channels sets off a flurry of nerve activity as cells send signals to the spinal cord about the change in temperature. The messages are then transmitted all the way to the brain, letting you know whether it is, in fact, getting hot in here. Extreme heat or cold not only stimulates the temperature-regulated channels but also the body's pain receptors.

If a warm or cool sensation is present more than momentarily, the thermal receptors gradually adapt, lessening or altogether stopping their response. (When you first stick your toe in the ocean, the water feels freezing. But by the time you've spent a few minutes in the surf, it feels comfortable.)

Speaking of adaptation, the hypothalamus controls sweating when your body gets too hot, which cools you as your sweat dries, and shivering when it gets too cold, which warms you right back up again.

Researchers are still learning a lot about these temperature-regulated channels. Curiously, they also play a role in our sensation of certain tastes. One subset of the channels that detect heat is also stimulated by capsaicin, an ingredient in hot peppers. And some of the channels that open in response to cold temperatures are also activated by menthol, a key ingredient in mint and minty flavors. Those gum commercials promising you minty cool breath are actually telling the truth.

COCKTAIL PARTY TIDBITS

❖ Eating something chilly can cause that all-too-familiar "brain freeze." This unpleasant sensation is caused by the rapid temperature change of the blood vessels along the roof of the mouth. The blood vessels overcompensate for the cooling and swell up as they frantically warm themselves. This expansion is detected by pain receptors and nerves in the face, causing an ice-cream headache.

❖ Certain animals, such as snakes, have receptors that can detect infrared radiation, or heat. This ability helps them find the hot little bodies of their rodent prey.

SMELL

Smell, or olfaction, was one of the first senses to evolve. When an airborne molecule enters the nose, it will encounter hundreds of different kinds of specialized odor receptors. An odor molecule and a receptor fit together like a lock and key, and the molecule will have to find its way to one of the few receptors it fits. The eventual binding of a molecule to a receptor generates an electrical signal, which is sent to the brain's olfactory bulb and then even deeper into the brain for processing.

But how does the brain make scents, er, sense, of these electrical signals? Identifying odors is a problem of pattern recognition. Imagine odor receptors are like piano keys. There are eighty-eight keys on a piano, but far more than eighty-eight sounds a pianist can produce at any given moment. Hold down three particular keys and you get one chord, change one of the keys and you get a different chord.

Odor receptors work the same way. A certain three receptors are activated, and the brain smells chocolate-chip cookies. A different set of five and the brain identifies the boss's cologne. It turns out that it's not really the "nose that knows"—it's the brain. But that's not nearly as catchy.

Odor is a large component of our perception of taste, and people with excellent palates—wine tasters or coffee connoisseurs, for instance—often have discriminating senses of smell.

Recent research illustrates that this expertise may be just a matter of exposure. In one study, scientists had human volunteers smell either a minty or floral odor for several minutes. This brief exposure made the subjects "experts" in related smells; those who sniffed the minty odor, for instance, were better able to differentiate among an array of minty scents.

It doesn't take much, it seems, for our brains to become adept at making the distinction between a cabernet and a merlot. Bring on the hundred-dollar bottles.

COCKTAIL PARTY TIDBITS

❖ Dogs are much better sniffers than humans. Canines have many more centimeters of olfactory tissue than humans do and a hundred times as many olfactory receptors per centimeter.

❖ But man's best friend can't lay claim to all the olfactory chops. In a recent experiment, scientists showed that college students were just as adept as dogs at tracking a chocolate scent across a field. The researchers say their results suggest that we humans haven't lost our ability to track scents—we're merely out of practice.

PHEROMONES

Pheromones are odorless molecules that are secreted by some animals to influence the behavior of other members of the same species. In mammals, these substances are detected by the vomeronasal organ, which is located somewhere between the mouth and the nose. Pheromones might be used to leave a trail, mark a territory, or attract a mate. Rats use pheromones to sniff out mates that are genetically different from them—a choice that could maximize the genetic fitness of their offspring.

Can humans detect pheromones? Research hasn't yet yielded a definitive answer, though it has provided some tantalizing clues. Women who spend a lot of time together have synchronized menstrual cycles, and pheromones are thought to be the mechanism behind this phenomenon.

There's also some evidence that pheromones influence human sexual preferences. Certain chemical substances we excrete—some say pheromones—can increase sexual arousal or desire in the opposite sex. Like rodents, women seem to be able to sniff out men who are genetic opposites. Researchers tested this by having women sniff a bunch of men's sweaty T-shirts. The ladies preferred the shirts worn by men who had immune systems that were genetically dissimilar to their own.

That's all well and good, but who wants her date to show up in a sweaty shirt?

Pheromones aren't only for finding sexual partners. A study of newborn rabbit pups revealed that a pheromone excreted in momma rabbit's milk facilitates learning. Rabbit pups engage in a complex sequence of behaviors when they prepare to nurse. These actions are stimulated by odors the pups associate with milk.

Over the course of their early lives, the pups quickly learn that certain smells mean that milk is near. Research has now shown that a pheromone in the milk itself aids this learning process. When scientists paired the pheromone with random, neutral scents, the rabbit pups learned to associate those scents with milk.

COCKTAIL PARTY TIDBITS

❖ Studies that examine the human capacity to detect pheromones often involve rubbing human sweat above volunteers' upper lips. (They'd better pay subjects handsomely for that.)

❖ Researchers have exploited insect pheromones to design new strategies for pest control. Chemicals that act as pheromones can confuse pests or reduce the number of insect matings.

TASTE

My favorite of the five senses. It's useful to divide taste into five basic component tastes: bitter, sweet, salty, sour, and umami (a Japanese word for a taste we perceive as "meaty" or "savory"). Technically, our tongues can detect other taste components, such as astringency, but most people refer to the five tastes listed above as the primary ones.

The tongue is covered with taste buds, each of which contains many taste cells. (Though most of our taste buds are on the tongue, there are others on the roof of the mouth, in the throat, and elsewhere.) Taste cells respond to chemicals in the food we eat, registering its component tastes, their concentration, and whether the overall effect is "good." The brain analyzes these signals, tells you that you're eating Brie, and makes your hand reach out for more.

Taste, however, is not a uniform experience. Some people have a higher sensitivity to salt or spice, for example, and a flavor that elicits a response of disgust from one brain might please another. There are also "supertasters" who perceive a number of tastes more intensely than the rest of us.

The sense of smell plays a critical role in our perception of taste. Need proof? Buy a bag of assorted jelly beans. Hold your nose and pop one in your mouth. What flavor is it? Now let go of your nose. *Mmm,* buttered-popcorn flavor.

Researchers have begun to unravel a strange phenomenon known as "thermal taste," in which changes in temperature alone can produce or alter the tastes we perceive. A recent study identified the mechanism that controls thermal taste, pointing to microscopic channels present in our taste buds. When a molecule of food hits our taste receptors, it prompts these channels to open, sending electrical messages to the brain. Temperature controls how wide the channels open, influencing how strong the signals sent to the brain are and how strongly we perceive a certain food's component tastes. (Warm ice cream will cause the channels to open up more and therefore taste sweeter than cold ice cream.)

COCKTAIL PARTY TIDBITS

- ❖ Taste sensitivity declines as we age. (Maybe that's why Grandma is always asking you to pass the salt.)
- ❖ Smoking also reduces taste sensitivity.
- ❖ Sour tastes cause an increase in the secretion of saliva—the body's attempt to dilute whatever's causing you to pucker.

HEARING

The process of hearing involves a series of mechanical movements, each of which causes another movement, until "Aha!"—we realize that the damned alarm is going off again.

The process starts when sound travels through the external ear to the eardrum, a thin membrane, and causes it to vibrate. This motion causes a bone called the hammer to quiver, which causes the same motion in a bone known as the anvil, which in turn does the same to an adjacent bone called the stirrup. In this way, the reverberations are carried all the way to the cochlea, a fluid-filled spiral tube that resembles a snail shell.

A sound vibration hitting the cochlea is like a stone hitting the surface of a lake—it causes ripples in the fluid. These ripples cause movement in some of the tens of thousands of so-called hair cells that fill the cochlea. The movement of these hair cells generates nerve signals that are carried by the auditory nerve to the brain.

Now it's up to the auditory cortices in the brain's temporal lobe to interpret the sound. Not only can the brain identify sounds in a split second, it can also detect tiny differences in when sound waves reach each ear, allowing us to determine where a sound is coming from.

Long before we're able to detect sounds, cells in our ears are generating their own, a recent study says. Researchers studied the developing auditory system in rats, which begin to hear when they're about two weeks old.

They discovered that, even before the rats are able to process sounds from the outside world, certain cells in their cochlea are buzzing with activity, firing off signals that eventually reach the brain. The scientists suggest that this self-generated sound helps ensure that the auditory system in the developing brain establishes the proper connections and wiring. (A similar system is believed to be at work in humans. But since we're able to hear from birth, this process would take place when we're still in utero.)

COCKTAIL PARTY TIDBITS

❖ The hammer, anvil, and stirrup are the three smallest bones in the human body. The stirrup is the smallest, one-tenth of an inch long.

❖ Dolphins and whales, which use sound to communicate with one another and locate their prey, have exquisitely sensitive auditory systems. They can hear sounds over a wide range of frequencies and long distances; whale songs are capable of traveling thousands of kilometers. (Researchers say, however, that because of noise pollution in the world's oceans, the range of these songs is shrinking.)

VISION

Vision is all about light. When we talk about "seeing" something we're actually talking about how our eyes respond to light, which enters the eye through the pupil and is focused by the cornea and the lens along the way. The light is then projected onto the retina, a layer of cells on the back of the eye.

The retina contains photoreceptors known as rods and cones. Cones help us see in bright light and are responsible for our ability to see color. Although cones respond to only three different colors—red, blue, and green—different patterns of cone activation allow us to distinguish among millions of different shades. Rods, on the other hand, do not detect color and work best when light is dim. (Oversimplified rule of thumb: We use cones to see during the day and rods to see at night.) Rods and cones use the optic nerve to send signals to the brain; the information is ultimately processed in the visual cortex.

Humans have binocular vision, which just means we use two eyes to see. This enlarges our total field of view and gives us the ability of depth perception. Go blind in one eye and you'll still be able to do many things—but you'll have to abandon your dreams of playing major league baseball.

Imagine a world without color. This is the world of people with achromatopsia, a genetic mutation that results in defective cone cells. Scientists have identified a promising course of gene therapy that might reverse this affliction.

The procedure involves inserting a normal copy of the relevant gene into a harmless virus. In studies, scientists injected this virus into the eyes of mice with achromatopsia. The healthy gene helped restore function to the cones, giving normal vision to the formerly color-blind mice. Perhaps now they'll be able to avoid that wretched farmer's wife.

COCKTAIL PARTY TIDBITS

- The image projected onto the retina is actually upside down. It gets turned right side up again by the brain. In one experiment, subjects were given glasses that flipped the world upside down. But their brains adjusted, flipping the image right side up again. When the glasses came off, the subjects saw the world as topsy-turvy for days.
- Color blindness is more common in men than women. The most common form is the inability to distinguish between red and green.
- Most mammals today are red–green colorblind. In fact, our shared mammalian ancestor was unable to see the wavelengths of light that we characterize as "red." Genes that allowed the visual perception of the color red evolved in apes and monkeys and were passed down to us.

SYNESTHESIA

We all know that a saxophone is a musical instrument, built to create sound. If you go to a jazz club, and the saxophonist steps forward for his solo, you prepare to hear music rather than, say, taste steak.

But what if, whenever the saxophonist played a D, you not only heard a note but saw the color blue? And whenever he played a B-flat, you heard the sound *and* saw the color red? For most of us, this experience would be pretty strange, but for the small percentage of people with synesthesia, these kinds of experiences are ordinary.

Synesthesia is a sensory mashup in which one sense consistently evokes another. The experience is variable and can involve nearly any combination of senses. People with the condition might see sounds ("The B-flat looks red") or taste words ("The word *bicycle* tastes like peas").

Synesthesia runs in families, but the genetic underpinnings haven't yet been identified. The condition is not considered to be a disease or disorder—most people report liking the experience. (And given synesthesia's resemblence to a hallucinogenic trip, minus the hallucinogens, who can blame them?) Indeed, many synesthetes, as those with the condition are called, take years to realize that their sensory cross-wiring is even unusual.

We may all be born with the capability to experience sensory mixing. The brains of newborns are dense and tangled with many connections among different structures and areas. For instance, babies might have unnecessary links between their auditory and visual systems. If they develop normally, this extraneous connection will eventually be pruned. But if the trimming process goes awry, the connection between sights and sounds might persist into adulthood. In such cases, sounds might activate both the auditory and visual centers in the brain, causing synesthesia.

COCKTAIL PARTY TIDBITS

- ❖ Synesthesia is more common among artists and creative types than the general population. Famous synesthetes include Duke Ellington, David Hockney, and Vladimir Nabokov.

- ❖ V. S. Ramachandran, one of the leading researchers of synesthesia, has suggested that studying the brains of synesthetes might reveal the neurological basis of metaphors. For instance, we might speak of a T-shirt as "loud" or a piece of cheese as "sharp." The human capacity for making such metaphors may be a result of the same processes that give rise to synesthesia.

STRESS

A stressor is anything that threatens your body or its resources, and stress is the state of being that results. But you know about stress. It's what you lug to work every day, along with your resentment and despair.

Stressors activate the hypothalamus, which triggers reactions throughout the body. Your adrenal gland releases adrenaline, your heart pounds, your breathing speeds up—all of which prepare you for "fight or flight." Meanwhile, as certain muscles and body parts ready themselves for action, nonessential functions like digestion slow down, allowing you to direct more resources toward the threat.

The stress response evolved primarily to deal with physical threats—predators, heat, hunger—but in the modern world, stress is often mental. That performance review or blind date is triggering the same changes that evolved to help your ancestors flee from a charging hyena. That's why the stress response sometimes feels counterproductive; it developed in a world very different than our own.

Chronic stress can take quite a toll on the brain and body. The adrenal gland also releases the hormone cortisol in times of stress. Over the long term, elevated cortisol levels can suppress the immune system, slow wound healing, and even increase abdominal fat.

A number of studies have revealed that stress hormones can do some truly terrifying things. Mice given cortisol for more than two weeks developed symptoms of anxiety.

Another study of mice showed that giving the rodents doses of stress hormone–like substances for a week accelerated the formation of brain abnormalities that have been linked to Alzheimer's disease. Norepinephrine, one of the body's stress hormones, can stimulate the growth of malignant tumors. So take off your shoes, uncork some wine, and relax.

COCKTAIL PARTY TIDBITS

- In the 1960s, scientists developed a scale that ranks events according to how stressful they are. The most stressful event? Death of a spouse. According to the scale, marriage is slightly more stressful than being fired.
- Guess what else is stressful? Switching to a low-fat diet. A study of mice forced to limit their fat intake revealed that the rodents had elevated levels of a stress hormone in their brains.
- One-third of people report having sex to relieve stress, according to a survey. Two-thirds say they listen to music. Sounds like an opportunity to multitask, if you ask me.

GOING MENTAL: HIGHER FUNCTIONS

LEARNING AND MEMORY

Memory makes us who we are. Short-term memory, also known as working memory, allows us to remember things long enough to complete immediate tasks. It's how we remember a phone number between the moment we look it up and the moment we dial.

Most experts say short-term memory can be as short as thirty seconds (though some include anything remembered for up to several minutes). Either way, it's not a long time. So in order to have a fighting chance at actually remembering things in the next hour, day, or month, memories need to move from short-term to long-term storage.

As scientists understand it, the mechanism for long-term learning involves changes to the strength of connections between neurons. The more you repeat a list of vocabulary words to yourself, for instance, or try to walk on stilts, the stronger the connections between certain neurons get.

It's a little like weightlifting—the synapses you use get stronger, while those you neglect grow flabby and weak. A process known as long-term potentiation helps these changes become lasting ones. And thus, a memory is born. The thalamus, hypothalamus, and brain structures known as the mammillary bodies are thought to be especially important in creating long-term memories, most of which eventually end up in the cerebral cortex.

An afternoon nap can speed the process by which short-term memories become long-term ones, a process known as memory consolidation. A recent study revealed that a ninety-minute snooze in the afternoon helped volunteers learn a new motor skill faster than those who stayed awake. Scientists still aren't sure exactly how sleep aids consolidation, but they hope that further research will perhaps allow them to artificially recreate sleep's benefits. Until then: Stop feeling guilty about that siesta.

COCKTAIL PARTY TIDBITS

❖ Our short-term memory capacity is limited. Scientists have actually put a number on it: We can hold seven discrete items in our short-term memory, plus or minus two. That's one reason it's convenient that phone numbers are only seven digits long. Now go get that guy's number.

❖ Scientists used to think that memories were stored in neurons themselves, with a single neuron equaling a single memory. It might sound silly to us now, but when you really think about them, so do a lot of current theories about the brain.

TYPES OF MEMORY

There is no single memory center in the brain, in part because there is no one kind of memory. The most important division in memory is the division between "declarative" and "nondeclarative" memories.

Declarative memories are those things we consciously recall when we need them: facts and personal experiences. Storage of these memories depends largely on the hippocampus and and parts of the cerebral cortex.

Nondeclarative memories, on the other hand, are largely unconscious—they are things we have learned but can draw upon without having to think about it. Take riding a bike. This skill is something you had to learn. But once you did, it became stored as a nondeclarative memory. Now, when you hop on a bike, you don't need to consciously remember how to pedal or balance. Your body just does it. Memories like these are stored with the help of the cerebellum, which helps coordinate movement, and the basal ganglia.

Other brain systems participate depending on the content of the memory. If it has an emotional component, the amygdala gets involved. Memories of things you've seen involve the storage of visual and spatial information, and things you've heard travel through a memory loop specifically for sound processing. Of course, the way we actually use our memories rarely obeys these tidy distinctions.

Scientists have discovered a new type of cell in the brain that they believe may be responsible for our memory of smells. The neurons are located in the olfactory bulb, part of the brain that processes odor.

Unlike other neurons in the olfactory bulb, the newly discovered cells keep firing even after we've smelled something, a property that seems to be responsible for our short-term memories of odors.

COCKTAIL PARTY TIDBITS

❖ Vivid memories of unexpected but significant events (such as where you were when you met your sweetheart) are known as "flashbulb memories" and usually contain great detail. Having a flashbulb memory, however, is not the same as having a photographic memory, which researchers believe may not really exist.

❖ You may curse when you forget to pick up your kid from preschool, but forgetting is healthy. There are cases of people who can't forget and are often burdened with memories of an overwhelming number of unimportant facts and details.

FALSE MEMORIES

THE BASICS

On January 26, 1986, the space shuttle *Challenger* blew apart about a minute after taking off, killing all seven people aboard. It was a vivid and highly memorable event, so the next day two researchers at Emory University decided to use the opportunity to conduct a study of memory.

They began by asking their students to write down how they heard the news. Several years later, they asked the same students the same question and compared those accounts with the first versions. The latter accounts were riddled with errors, but the students were highly confident in their erroneous statements. It was a vivid demonstration of what scientists now know well: Our memories are far from infallible.

Researchers have used a variety of techniques to "implant" erroneous memories in the minds of volunteers. For instance, they have asked people to imagine, at length, an event that never happened. Surprisingly, when asked about the nonexistent event later, a significant proportion of participants believed it did happen after all.

Controversy has also swirled around "recovered memories" of childhood abuse. Skeptics suggest that such memories, which often appear during psychotherapy, may be products of suggestion by a therapist. (Many so-called repressed memories are recovered during hypnosis, a process that makes people even more vulnerable than usual to the power of suggestion.) Though false memories surely exist, it is extremely difficult to determine the veracity of any one particular memory without corroborating proof.

ON THE FRONTIER

One recent study showed that it may be possible to distinguish real memories from false ones. Researchers implanted false memories in volunteers. Then they used functional MRI machines to scan participants' brains while they recalled both false and real memories.

The scientists found that the medial temporal lobe, which is important in remembering specific details, was active when participants were recalling real memories but not when they were "recalling" false ones—even though the volunteers were confident that the latter memories were real.

COCKTAIL PARTY TIDBITS

❖ Patients who believe that their shrinks have implanted false memories have sued their doctors for malpractice and won.

❖ During the '70s and '80s, a rash of high-profile cases emerged in which people said they had been victims of UFO abductions or satanic cults. Many of these claims were later attributed to false memories.

EMOTION

Emotions involve both our minds and our bodies (think of the racing heart that accompanies fear), but scientists still aren't sure how the two interact to create deep feelings. A brain network known as the limbic system is at the center of it all, an ancient part of the brain that we humans share with many of our ancestors. Some of the most notable components include the amygdala, involved in fear and reward; the hippocampus, involved in memory; and the hypothalamus, which regulates the body's automatic processes.

The emotions we experience can be considered combinations of a smaller number of primary emotions. Scientists can't agree on how many of these basic emotions there are, but they at least include the big four: fear, anger, sadness, and happiness. Surprise and disgust are sometimes also added to this list.

Emotions are instinctual—even infants express them—and they're also universal. Scientists have shown faces expressing anger, fear, surprise, and other emotions to people around the world. No matter what culture the participants belong to, they agree that the scowling face is angry, the face with raised eyebrows and a dropped jaw is surprised, and so on. Of course, culture also exerts a huge influence on when emotions are expressed. The food that disgusts you might be a delicacy on the other side of the world—or just to those who frequent the truck stop down the highway.

ON THE FRONTIER

Think of some of the events in your life that you recall most vividly. Your first, fumbling kiss? Almost crashing your new car? Whatever the memories, they probably involve strong emotions. Indeed, events that evoke intense emotion tend to be remembered most clearly.

One recent study found that norepinephrine, a stress hormone released when we experience strong emotions, strengthens the connections between neurons, potentially explaining why memories of emotional events are so sharp.

COCKTAIL PARTY TIDBITS

❖ One of the most intriguing disorders of the mind is called Capgras syndrome. Patients with Capgras insist that their loved ones— spouses, parents, children—are actually identical imposters. These patients recognize their family members visually, but for some reason do not experience the emotional arousal that normally accompanies the sight. So, the patient thinks, *This woman looks just like my wife, but I don't feel any emotion when I look at her, so it must be an impostor.* And they called *Invasion of the Body Snatchers* science fiction!

FEAR

When it comes to processing fear, multiple brain circuits seem to be operating simultaneously. The amygdala acts as the brain's fear center. Imagine watching a snake through a glass display at a zoo. If the snake suddenly rears up and strikes at the glass, the amygdala's alarm bells start going off.

The amygdala sends messages to the hypothalamus, which triggers the stress response, readying your body for fight, flight, or crying like a baby. (This part of the fear response is the same as the stress response we discussed a few sections back.) Meanwhile, the amygdala sends messages to the frontal cortex, which gets busy assessing the threat, determining whether you really need to run from the snake. This system operates slightly more slowly, however, which explains why your heart will start pounding before—and keep pounding after—your brain figures out that you're protected by a sheet of glass. The brain will sometimes cause the body to respond to nonexistent threats, but if you're facing a real danger in which every second counts, you'll be glad your heart started racing right away.

Anxiety is a variation on fear. Whereas the fear response occurs immediately upon experiencing something frightening and doesn't last very long, anxiety is a more general sense of dread; it can begin in anticipation of something scary and last a long time.

The human brain is wired to be particularly responsive to human faces, especially those wearing expressions of fear. Research reveals that we process images of fearful faces faster than those that are happy or neutral.

Scientists say this speed makes sense: Seeing a fearful expression on someone else signals something dangerous on the horizon and means we'd better get our own butts in gear if we want to survive. From an evolutionary perspective, it's not nearly as important to process the happiness of others.

COCKTAIL PARTY TIDBITS

❖ Speaking of threatening faces: People with higher anxiety levels are more likely to notice anxiety in others.

❖ Disable a rat's amygdala and he'll befriend his sworn enemy: the cat. Experiments show that destroying the structure leads rodents to approach a sedated cat located nearby. Some rats will even crawl on top of a feline and nibble its ear. (Love bites?) Needless to say, rats with functioning amygdalas do not exhibit this behavior.

HAPPINESS

Implant an electrode in the right part of a rat's limbic system and he'll be the happiest rodent around. If the rat is given a lever that stimulates this electrode, he'll press it thousands of times an hour, instead of even eating. Certain areas of the limbic system, it turns out, are part of the brain's pleasure network, and stimulating them creates a natural high. The brain has several different structures that generate feelings of well-being—the nucleus accumbens is a big one—and they all tend to use the neurotransmitter dopamine to make you feel good. These brain circuits all seem to be involved in what we label "happiness."

Researchers have conducted global studies of happiness. These studies usually measure "self-reported well-being," which is a fancy way of saying that they ask people how happy they feel. The results may surprise you.

Happiness, it turns out, is evenly distributed across nearly all demographics—class, sex, age, race, educational level, and more. But happy people have certain traits in common. The contented among us tend to have high self-esteem, feel in control of their lives, be optimistic, and be extroverted. Of course, it's difficult to untangle cause and effect here—are these people happy because they feel in control or do they feel in control because they are happy?

The former Duchess of Windsor once declared that it's impossible to be too rich or too thin. But is it possible to be too happy? Researchers have recently published what they say is the first study of whether there is an optimal level of happiness. They discovered that people who reported the highest levels of happiness were less educated and less politically involved than those who said they were merely moderately happy. The happiest people may not always be motivated to try new things, achieve more, or change their lives. It *feels* good, but happiness, it appears, can breed complacency.

COCKTAIL PARTY TIDBITS

- ❖ **The happiest place on Earth? Denmark, according to surveys showing that residents of that country rate themselves as being the most content. Sorry, Disney World.**
- ❖ **Marriage does boost happiness, but only very slightly, research shows.**
- ❖ **It turns out that it's not the absolute amount of money you have that makes you happy—it's your financial standing compared to your friends, neighbors, and peers. Stay a step ahead of the Joneses, and you'll feel just fine.**

HUMOR AND LAUGHTER

As anyone who has had a joke greeted by a room of stunned silence knows, senses of humor can vary greatly and are undoubtedly influenced by culture. But researchers believe that some universal structure or trait must underlie humor— why do we all have the capacity to be amused by a made-up story or ridiculous scenario?

Theories about humor vary: Maybe jokes are funny because they make us feel superior to that idiot who walked into a bar. Or maybe they surprise us: we expect one outcome and get another. Or maybe they make some of our deepest fears (death, for instance) nonthreatening. Humor always has one or several of these characteristics, but no overarching theory of what makes things funny has been universally accepted.

Laughter is an obvious outcome of humor, but it can occur even without punch lines. An observational study of social interactions on a college campus revealed that speakers more likely to laugh than listeners, and that the majority of laughter was in response to utterly conventional statements (things like "See you later") and not to actual jokes. These findings are consistent with the idea that laughter evolved to help us bond with one another.

In other words, watching *The 40-Year-Old Virgin* just isn't as funny if you're actually a 40-year-old virgin.

Jaak Panksepp, a researcher known for his work on animal emotion, play, and pleasure, says that even rats laugh. How does he know? He tickled them. In response, the rats consistently made a chirping sound that Panksepp identifies as rodent laughter. Some have challenged the use of the word *laughter,* to describe what the rats are doing, but Panksepp is amassing evidence that their reaction is not very different from what we humans do when we crack up. One thing's for sure: The idea of a man with a Ph.D. systematically tickling rats is very funny.

COCKTAIL PARTY TIDBITS

- ❖ Laughing boosts the immune system and lowers the body's level of stress hormones.
- ❖ If it's not quite possible to die of laughter, it is possible for laughter to signal impending death. There are a handful of cases in which a seemingly healthy person starts laughing uncontrollably, only to die within a day or two. Laughter, it turns out, can be a symptom of damage to the limbic system, such as that caused by a brain hemorrhage.

LOVE AND ATTACHMENT

Love can actually be broken down into three components: lust, attraction, and attachment. Lust, or the desire for someone, anyone, is most clearly regulated by the sex hormones: testosterone and estrogen (even in women, testosterone increases sex drive).

Attraction, or the desire for one special someone in particular, appears to be a product of several different substances, including the neurotransmitter dopamine, which seems to turn up whenever the brain is processing good feelings. Staring at a photo of one's beloved can activate the dopamine-regulated brain regions involved in pleasure and reward.

The most mysterious part of romantic love is that third part: attachment. Humans are in the minority of species that form monogamous bonds. (Yep, most animals are even more promiscuous than we are.) But studies of animals that do form strong attachments with their mates—like the prairie vole—have revealed much about the neuroscience of bonding.

In particular, the research points to the importance of oxytocin, a hormone released during both orgasm and breastfeeding. Oxytocin has been shown to increase trust in men and women alike. Injecting male prairie voles with oxytocin prompts them to spend more time around their mates, and blocking its activity leads the males to fornicate and flee.

Call them crazy, but scientists say that the love a mother feels for her newborn is not all that different from the head-over-heels love you felt for your high school sweetheart. Studies of mothers have shown that their feelings for their children involve activation of many of the same pleasure-provoking brain circuits that participate in romantic love.

What's more, both kinds of love dampen parts of the brain we use to make negative social judgments about others. At last, your friend's devotion to her creepy, socially inept fiancé explained.

COCKTAIL PARTY TIDBITS

- Why we kiss is not entirely clear—from an evolutionary standpoint, it isn't necessary—but some have proposed that it originated in the act of female primates feeding their offspring by mouth. This lip-on-lip action may have evolved into a sign of affection.

- Losing a loved one really can break your heart. The trauma of a sudden emotional shock, such as the death of a spouse, can cause a heart attack. Doctors have named the condition "broken-heart syndrome."

CREATIVITY

Though we usually think of creativity as a hallowed trait, possessed only by great geniuses, we all use creativity, which at its base involves the generation of novel and useful ideas in our everyday lives.

There is no special brain center for creativity; indeed, some researchers have suggested that creativity results from connectivity among different parts of the brain. But the frontal and temporal regions of the cerebral cortex in particular turn up as important areas for creative ability.

Especially creative people seem to have certain characteristics in common, including a high level of general intelligence, expert knowledge about one or more subject areas, flexible patterns of thinking, and a high tolerance for risk. There is also a definite relationship between creativity and mental illness, particularly mood disorders, which are unusually prevalent among eminent artists, writers, musicians, and others. The connection hasn't been totally unraveled yet—it's unclear whether mood problems make people engage in creative endeavors or whether some quirk of brain chemistry causes both mood problems and creativity.

On the other hand, studies have also revealed the healing power of creativity, and such art forms as music, art, and writing can boost immune function and promote psychological well-being. Eureka.

ON THE FRONTIER

If you want to tap into your creative abilities, tell that critic inside to stop pestering you. That's the lesson from a study of jazz musicians. The musicians had their brains scanned while they played compositions they had memorized and then again while they riffed on these compositions.

The scans revealed that when the musicians were improvising, activity decreased in the parts of the brain usually involved in judging and evaluating our own actions. The finding suggests that it might be important to silence self-criticism for your ideas to flow freely.

COCKTAIL PARTY TIDBITS

- Studies of people who work in the creative arts reveal that they are eighteen times more likely to commit suicide, eight to ten times more likely to experience depression, and ten to twenty times more likely to suffer from bipolar disorder than the general population.

- Brain injuries have occasionally been known to drastically ramp up someone's creativity. One man became an avid painter and sculptor after a brain hemorrhage. And an elderly woman saw her artistic skills improve as dementia ravaged regions of her frontal and temporal lobes.

LANGUAGE

Aside from what you've seen in the movie *Doctor Doolittle*, experts still disagree on whether animals are capable of language as we define it. We do know, however, that some of the animal kingdom's geniuses can perform some pretty astounding linguistic feats. Primates have proven themselves able to learn signs from researchers who teach them a modified kind of sign language (since they can't articulate words the way humans can). These animals can use signs to communicate their needs to researchers, and have also been known to invent new signs (or "words"). That's all well and good, but when are monkeys going to learn to type out one of Shakespeare's plays?

In us lucky *Homo sapiens*, many different brain areas are involved in language, but two in the cerebral cortex are particularly important. Wernicke's area is involved mainly in understanding language and the meanings of words. The second, Broca's area, helps control grammar and the production of speech. Language disorders, or aphasias, dramatically illustrate the difference between Broca's and Wernicke's areas.

Someone with damage to Wernicke's area will speak in what sounds like fluent sentences but use gibberish words instead of real ones. Someone with this kind of aphasia might say, for instance, "Well, I went to aronize my, you know, the filibart. It was ringering my toast." The words might not make any sense, but the sentence has a grammatical structure. On the other hand, people with damage to Broca's area will understand the meaning of words but be unable to string them together in any meaningful way.

Gestures are a vital component of language. A recent study showed that using hand gestures while learning something new improves your ability to remember it later.

Scientists had third- and fourth-graders learn simple algebraic concepts. Some children learned gestures that accompanied the concepts and were encouraged to use these gestures when solving problems on their own. Several weeks later, the students who learned and used gestures as a problem-solving strategy were far more likely to remember the concepts.

COCKTAIL PARTY TIDBITS

- There are some six thousand languages spoken in the world today. Hundreds of them are rapidly dying out.
- In the nineteenth century, the Linguistic Society of Paris created an outright ban on discussing the origin of language. Today, the evolutionary origin of language is a hot research topic.

LANGUAGE ACQUISITION

From "mama" to "I want another cookie!" children learn language in predictable stages. Babies are babbling away by the time they're about six months old, using words by age one to eighteen months, and speaking in sentences by the time they're two.

What children hear is vitally important to the development of language skills. For instance, research generally shows that babies need to hear live speakers—television dialogue alone isn't enough to aid language development.

Some linguists believe that we come into the world hardwired for language, that we're born with a "universal grammar," or set of mental rules we can apply to whatever language we grow up hearing. Others contend that the patterns and rules for language are almost entirely learned by listening to lots of talk as babies.

Many of the world's most renowned linguists have tackled some version of the question: If a group of infants were allowed to grow up on a desert island all by themselves, would they develop anything resembling language? The experts' responses run the gamut, and we may never know the real answer. On the other hand, if there are any producers in search of a new concept for a reality TV show . . .

Preschoolers learning a second language go through the same stages as infants learning their native language. Researchers studied Chinese toddlers who were adopted into American families.

As they were exposed to English, the adopted children started by picking up many single nouns, eventually producing short pairings of nouns and verbs ("Ball fell"). These are the same steps an infant will go through learning a language.

COCKTAIL PARTY TIDBITS

❖ In one enterprising study, researchers tape-recorded parents' interactions with their babies over several years. Researchers found that the more parents talked to their kids, the higher their childhood IQs were. The amount of time babies spent listening to their parents talk was an even better predictor of future smarts than socioeconomic status.

❖ Different languages are comprised of different sounds, or phonemes. From the moment babies are born, they are capable of discriminating between all the phonemes that exist in any language. But by about six months of age, they start to tune in exclusively to the sounds that exist in the language (or languages) they regularly hear. Gradually, they lose the ability to distinguish other phonemes.

DECISION-MAKING

Chocolate or vanilla? Boxers or briefs? Have kids now or devote everything to your career first? Every day, we're faced with countless decisions that need making, some of them minor, some of them fraught with existential drama.

Decision-making is closely tied to inhibition and self-control. It is inhibition that allows us to make choices that delay gratification—for instance, deciding not to have a martini at lunch because we need to ace an afternoon presentation. Inhibition and self-control are governed by the prefrontal cortex. Damage to the frontal lobe can impair decision-making abilities; people with frontal lobe injuries consistently chose a strategy that provided small and immediate rewards over one with bigger payoffs in the long term.

Decisions are not always based on reason alone, of course, but on a combination of reason and emotion. The amygdala helps evaluate the emotional relevance of a decision, and strong emotional responses can lead us to make choices that aren't exactly logical. Two kinds of decisions in particular, moral and economic, highlight the complicated interplay between emotion and reason that occurs when we're faced with tough decisions. We'll discuss them more in the following pages.

Much recent research on decision-making revolves around serotonin. In one study, researchers manipulated participants' levels of this neurotransmitter and then asked them to play a game known as the Prisoner's Dilemma, which involves cooperation and social decision-making.

In the game, players must decide how to divide a sum of money between themselves and the other participants. Those with reduced serotonin levels were less likely to cooperate or share the money fairly. This suggests that serotonin makes cooperation rewarding.

COCKTAIL PARTY TIDBITS

❖ Research shows that sleep deprivation impairs our ability to make sound decisions, even to the degree that drunkenness can. Probably best to get a good night's sleep before that blind date.

❖ How many decisions do you make about food each day? No matter what number you just guessed, it's probably wrong. In a recent study, subjects estimated that they made fifteen decisions about food a day. But when researchers probed further, they discovered that the participants actually made an average of more than two hundred food-related decisions daily.

ECONOMIC DECISION-MAKING

Capitalism is based on the idea that, when left to their own devices, individuals act in their own self-interest. However, recent research in the rapidly growing discipline of "neuro-economics" illustrates that humans don't always make choices that serve them well.

Imagine you are offered a 50 percent chance of winning $150 and a 50 percent chance of losing $100. Even though it's to your advantage to take the gamble—over the long run, you'd come out ahead—if you're like most people you'll refuse. This behavior is known as loss aversion, and neuroscientists have found that the brain is, indeed, more sensitive to losses than gains.

Ironically, people with damage to the emotional centers of their brains are far more likely to accept the gamble and end up making considerably more money than those with healthy brains. (Note: Employ a brain-damaged investment adviser at your own risk.)

In the gamble described above, participants knew their chances of winning. But in the real world, this is rarely the case—investors in the stock market rarely know their precise odds of coming out ahead. Recent studies have shown that people prefer gambles in which their odds of winning are known, and there is more activation in the amygdala, which processes fear, when the odds are ambiguous.

Naturally, businesses are interested in neuroeconomic findings. Of particular use may be a recent study showing that brain scans can predict when customers are going to buy a product. Researchers put volunteers in a functional MRI machine and scanned their brains while displaying the images and prices of products. Spikes of activity in the nucleus accumbens, the brain center that anticipates rewards, and part of the frontal cortex, which is involved in balancing gains and rewards, meant that the participant was likely to buy the product.

COCKTAIL PARTY TIDBITS

- ❖ Neuroeconomists have also studied whether monkeys can be taught to use money. Turns out they can—capuchin monkeys used tokens to "pay for" their favorite treats, like grapes. One researcher even witnessed a male capuchin giving a female his token in exchange for sex. And to think people still deny that humans and monkeys evolved from a common ancestor.
- ❖ More monkey business: The capuchins were also seen stealing "currency" from experimenters.

MORAL DECISION-MAKING

An out-of-control train is hurtling down the tracks, headed toward five railway workers, unaware of onrushing doom. You are standing near a switch—flipping it would send the train onto another track, saving the five workers but killing one man working on the second track. Would you flip the switch?

What if, instead, you were standing on a bridge above the train and could save the five workers by pushing a man onto the tracks? Would you push? Most people say they're willing to flip the switch to save the five workers, but not to push the man off the bridge. But in both cases, the math is identical: You're saving five lives by taking one.

This famous scenario, and others, are helping scientists discover the neurological basis of morality. What they are discovering is probably what you felt when imagining the above circumstances—that your reasoning pulls you one way while your emotion pulls you another. Considering flipping the switch activates parts of the prefrontal cortex that are associated with cold, hard logic and utilitarian choices. But imagining pushing a man off the bridge (a much more personal action) activates the emotional processing centers of the brain. This strong emotional response, researchers believe, trumps logical reasoning and explains seemingly illogical moral decisions. And even if we do decide to push that man off the bridge, it takes us a long time, neurologically speaking, to decide to do so. Lucky for him.

ON THE FRONTIER

Even babies have some sense of right and wrong, a new study suggests. Scientists used wooden dolls to present several scenarios to infants. In the scenarios, a climber figure tries to make its way up a hill. In one variation, a triangular-shaped figure helps the climber up the hill. In another, a square-shaped figure blocks the climber's progress.

When the figures were later presented to the infants, the babies almost universally chose to reach out to the helpful figure, suggesting that babies prefer good guys.

COCKTAIL PARTY TIDBITS

❖ People say they are more likely to commit minor moral infractions if they think everyone else commits them, too. (This explains speeding.) People are also more likely to violate moral codes if there's no clear human victim.

❖ Studies reveal that people behave less morally than they think they do. Researchers determined that people accurately predict how much money other people will donate to charity but overestimate their own willingness to contribute.

YOU BREAK IT, YOU BUY IT: PROBLEMS IN THE BRAIN

AUTISM

Temple Grandin, one of the most famous people with autism, said that the disorder left her so puzzled by human social interactions that she felt like she was an anthropologist from Mars. Indeed, autism is a developmental disorder mostly involving social and communication deficits. Autistic children may be very late to develop language skills, have difficulty interacting with others, become obsessive about hobbies, and acquire repetitive behaviors.

People who work with autistic kids say that if you have ten different autistic kids in a room, you have ten different autisms. Severe autism can leave children severely mentally retarded and isolated. At the other end of the spectrum is Asperger's syndrome, a high-functioning autism. These children often join their peers in regular classrooms and have above-average intelligence. Either way, autistic children thrive on predictability and routine.

The cause of autism is one of the mostly hotly debated and intensely researched subjects in neuroscience today. Many current theories revolve around the role of the immune system, and research has shown that exposure to viruses during pregnancy can increase a woman's chances of having a child with autism. Some parents of autistic children insist that a compound contained in common childhood vaccines caused their children to develop autism, but scientific studies have not borne this theory out. There has been a rapid rise in the number of cases of autism diagnosed in the last few decades—the cause of this increase is still a mystery.

Autistic children have abnormally large brains and fast rates of brain growth during the first year or two of life. Researchers still aren't quite sure what to make of these findings, but one of the most recent investigations suggests that rapid head growth during a child's first year might be useful for predicting autism even before more pronounced behavioral symptoms become apparent. A big or fast-growing head alone doesn't mean a child is autistic, but parents who notice it in their children should be on the lookout for other signs of the disorder.

❖ Though most autistic children have impaired cognitive function, many are also known as "savants." They perform exceedingly well on one or several highly specific tasks. They may be able to tell you the day of the week that corresponds to any date in history, perform huge calculations instantaneously in their heads, or draw accurate depictions of scenes they have seen in passing.

EPILEPSY

Epilepsy is characterized by repeated, unpredictable seizures. It affects children and adults alike (nearly one-third of those with epilepsy are children). There are many different kinds of epilepsy, and seizures come in all shapes and sizes. The ones you're probably imagining are accompanied by involuntary jerks and spasms of the body, but during others, the body remains rigid and still. Some seizures affect the entire brain, while others are more localized. Some, but not all, cause a loss of consciousness.

Before certain kinds of seizures, some epileptics report smelling strange odors, like the smell of something burning, or feeling certain unprovoked emotions, and these patients learn to recognize the warning signs of an oncoming attack. The cause of epilepsy is unknown, but seizures themselves are caused by a sudden flurry of uncontrollable neural firing in the brain.

There is usually no cure for epilepsy, though treatment—often in the form of medication—can reduce the frequency and severity of seizures. Epileptics have also had the honor of being subjected to some of the most radical forms of brain surgery. Kids with severe epilepsy sometimes have an entire hemisphere of their brains removed, and others can have the connection between the two hemispheres of their brains (the corpus callosum) severed. These drastic measures some-times provide dramatic relief for patients suffering from se-vere, life-disrupting seizures.

Researchers are beginning to use a more advanced brain scanning technique, called magnetoencephalography (MEG), to monitor seizures and pinpoint their locations. MEGs mea-sure brain activity in real time, unlike EEG. By locating the source of seizures more precisely, MEGs allow doctors to re-move the brain tissue involved in epilepsy without unneces-sarily damaging any of the healthy surrounding areas.

COCKTAIL PARTY TIDBITS

- ❖ The ancient Greeks believed that epilepsy was a supernatural phenomenon caused by the gods. In the Middle Ages, epileptics were viewed as possessed by evil spirits and sometimes persecuted as witches.
- ❖ It's commonly believed that, if you see a man thrashing around in the throes of a seizure, you should put something in his mouth to keep him from swallowing his tongue. Untrue. Putting something in a seizing person's mouth can be dangerous, and it's physically impossible to swallow your own tongue.

ATTENTION-DEFICIT/HYPERACTIVITY DISORDER

Attention-deficit/hyperactivity disorder is the single most commonly diagnosed disorder of childhood. Kids with ADHD have difficulties with some combination of attention, hyperactivity, and impulsivity. We all knew kids like this. They were the ones who fidgeted at their desks and kept getting out of their chairs, who couldn't wait their turns during games and took toys out of other kids' hands, who came up to you during recess and bit you on the back (okay, maybe that was just a kid at my school). Whether these behaviors indicate a genuine case of ADHD depends on their severity and frequency, as judged by a qualified expert.

ADHD is thought to result from a combination of biological and environmental factors, including brain abnormalities and high levels of family stress. Fortunately, ADHD is eminently treatable for most children. Scientists have discovered that, ironically, Ritalin and other stimulants seem to settle many kids down and give them back their focus. But Ritalin has become a controversial drug, in no small part because of the ever-increasing number of children who seem to be taking it. ADHD is a real disorder that can derail the lives of those afflicted with it, but it is almost certainly overdiagnosed; as many as two-thirds of children who are "diagnosed" never undergo psychological or educational testing for the condition.

ON THE FRONTIER

Brain imaging studies are revealing some of the neurological mechanisms underlying ADHD. Researchers have already found that certain brain structures—the prefrontal cortex, the basal ganglia, and the cerebellum—are 5 to 10 percent smaller in children with the disorder. Other studies have shown that these children also have reduced blood flow in these areas and that the lower the blood flow is, the worse the symptoms tend to be. Finally, scientists have discovered that Ritalin increases blood flow to these structures.

COCKTAIL PARTY TIDBITS

❖ ADHD is much more commonly diagnosed in boys than in girls, possibly because the symptoms are less obvious in girls.

❖ The symptoms of ADHD tend to decrease as children age.

❖ According to some estimates, there was a fivefold increase in Ritalin use between 1990 and 2000. Even preschoolers are getting in on the action.

❖ Ninety percent of the Ritalin produced each year is consumed in the U.S.A.

DYSLEXIA

In order to read properly, we need our eyes to perceive the letters and the order in which they appear, and our ears to make sense of the sounds of those letters. It is not surprising then that children with dyslexia, a learning disability characterized by reading troubles, have difficulties with auditory processing, visual processing, or both. Symptoms vary, but dyslexics may have trouble spelling, sounding out words, or even pronouncing the sounds of a word in the right order.

Much of dyslexia is currently thought to stem from a problem processing the sounds of a language. Good readers are able to register quick changes in sounds between each letter of a word as it is said aloud—for instance, the shift between the sound of the letter *p* and the sound of the letter *a* in *panda*. Dyslexics, studies have shown, have trouble hearing this fast shift as well as distinguishing the sounds of similar letters such as *b* and *p*.

These tasks are performed by certain auditory neurons that seem to be underdeveloped in dyslexics. Some dyslexics may also have problems with the visual demands of reading and report that when they try to read, the letters of a word move around or the word seems to blur or shimmer. Some scientists believe that these symptoms result from deficiencies in the visual pathways that allow better readers to keep images stable as their eyes move across the page.

ON THE FRONTIER

Recently, a small group of researchers suggested that dyslexia does not result from the inability to properly and quickly match letters to sounds. Instead, they contend, dyslexics have trouble filtering out background noise (which happens to be pretty common in school classrooms). The theory suggests that the dyslexic brain has trouble ignoring irrelevant sounds, making it harder to pay attention to the subtle sounds of the words that do matter.

By the time you're reading this, there might be an even newer idea about what causes dyslexia. Welcome to the maddeningly fast-paced scientific enterprise.

COCKTAIL PARTY TIDBITS

- ❖ Dyslexia has no relationship to intelligence; there are both very smart dyslexics and not-so-smart ones.
- ❖ Dyslexia is less common in Italy than in the United States. Scientists believe this is because Italian is an especially phonetic language. That is, Italian words are spelled more like the way they sound than English words are.

DEPRESSION AND BIPOLAR DISORDER

THE BASICS

Depression is more than just sadness. In order to be diagnosed with depression, a patient has to have two or more weeks of a disruptively low mood, as well as pronounced changes in sleep, appetite, concentration, energy, and more. People who are depressed often lose interest in seeing friends and loved ones and no longer enjoy even the best things life has to offer (like two-for-one drink specials).

A depressed brain looks much like a brain that has been chronically stressed. There's a reduction in serotonin activity, a disruption in the regulation of norepinephrine, and high levels of cortisol. Periods of depression will usually lift on their own (though medication and psychotherapy can help) but can recur throughout a lifetime.

Add a dash of mania, however, and that depression turns into bipolar disorder. Mania is an extended period of abnormally elevated mood, and people with bipolar disorder have alternating episodes of mania and depression. During manic periods, people may sleep very little, have racing thoughts and speech, have inflated self-esteem and delusions of grandeur, and be particularly impulsive, blowing paychecks on shopping sprees or engaging in reckless sexual activity. Years can separate manic and depressive episodes.

An accumulating body of research suggests that there might be a link between depression and the creation of new neurons, or neurogenesis. The exact cause and effect relationship isn't clear, but new theories suggest that depression is characterized by a reduced rate of neurogenesis in the brain and that increasing the birth rate of new brain cells might help alleviate it. Among the evidence: Chronic stress, which has brain effects similar to depression, reduces neurogenesis. Antidepressants, on the other hand, increase neurogenesis, and studies have shown that blocking this effect keeps such medication from working.

COCKTAIL PARTY TIDBITS

❖ One in twenty Americans will experience depression during any given year. Depression is twice as common in women, but bipolar disorder strikes both sexes equally.

❖ Depression that regularly occurs during the winter is a type of Seasonal Affective Disorder (SAD), which has been linked to the shorter exposure to daylight that comes during the colder months. One of the most effective therapies involves staring into artificial daylight for an hour or more during the winter months.

ANXIETY DISORDERS

Anxiety disorders hijack the body's natural fear system, causing abnormal and irrational levels of worry and dread. Generalized anxiety disorder is a chronic state of tension, in which sufferers spend nearly all day, every day, in a state of apprehension. Panic disorder, on the other hand, is characterized by periods of normalcy alternating with sudden panic attacks. The attacks are usually brief, less than half an hour or so, but can be terrifying: The body's fight-or-flight response is triggered, causing rapid heart rate, sweating, and heavy breathing, and patients usually experience a sense of impending doom. Panic disorder can be profoundly disruptive, causing sufferers to avoid places, situations, and events that might trigger an attack. (Worry about having an attack can be even more problematic than a panic attack itself.)

Both generalized anxiety disorder and panic disorder are often accompanied by a host of physical symptoms, including headaches, gastrointestinal problems, and insomnia. Many people with these disorders first seek medical help for one of the accompanying physical problems, and diagnosing the underlying psychiatric issue can be difficult. As you might expect, anxiety disorders are thought to be the result of a hyperactive fear system. They are also associated with high levels of noradrenaline and glutamate and low levels of serotonin.

Women are much more likely to suffer from anxiety disorders (and mood disorders, now that we're on the subject) than men are. New research suggests this discrepancy might be a result of sex-based differences in the brain's serotonin system (low levels of serotonin have been implicated in anxiety and mood disorders alike). Now, PET scans show that women have more serotonin receptors in their brains and less of the protein that helps recycle serotonin for future use. Scientists aren't yet sure how these differences translate into increased risk for anxiety disorders, but if you're looking for a dissertation topic for your neuroscience Ph.D., you could do worse than follow up on that one.

COCKTAIL PARTY TIDBITS

❖ Anxiety isn't just a human affliction—five million dogs suffer from some form of separation anxiety. And lest you think it's the dogs who've gone crazy: One-third of people surveyed report leaving the television or radio on to help calm the pooch.

❖ Panic seems to spare the young. A study of middle-school girls showed that only those who had completed puberty had ever had a panic attack. One more reason to pine for childhood.

PHOBIAS

There's a whole alphabet of fears, from acrophobia (fear of heights) to zoophobia (fear of animals). Phobias involve an irrational fear of objects or situations that are unlikely to cause real harm. These fears can interfere with daily life, as phobic people go out of their way to try to avoid dogs, bees, elevators, and so on. Phobias are a subset of anxiety disorders and, like many mental illnesses, are probably the result of just the right cocktail of genes and environment.

One of the most common treatments for phobias is systematic desensitization. It involves exposing a patient to his worst fears one small step at a time. If, for instance, you have gephyrophobia (a fear of crossing bridges), the first step in therapy might involve looking at a photo of a bridge. Once you can do that without feeling your stomach lurch, you might stand outside and gaze upon a real bridge. And then maybe watch somebody cross it. Then, if all goes well, maybe even take a few tentative steps across the bridge.

The idea is to progressively teach patients how to soothe their anxieties in the presence of terrifying stimuli. Doctors have also discovered that virtual reality can help patients face their fears. (It is hard to fall off a nonexistent bridge.)

Desensitization therapy doesn't help everyone, and researchers are working hard to find effective pharmacological interventions for specific phobias. Studies have suggested that fears and phobias are overcome not by erasing bad or traumatic memories, but by creating new memories—ones in which the snake is docile and nonthreatening.

Scientists are now looking into developing drugs that treat phobias by boosting the activity of brain receptors involved in learning. Indeed, some preliminary trials have shown that such enhanced learning can reduce the amount of desensitization therapy needed.

COCKTAIL PARTY TIDBITS

❖ Ever met anyone with an intense fear of flowers? Didn't think so. In a famous experiment, psychologist Martin Seligman demonstrated that it was much easier to teach humans to develop a phobia of snakes than a fear of flowers. The study plays into the theory that our brains are predisposed to develop certain fears.

❖ There's no better crowd-pleaser than trotting out the name of a strange-sounding phobia. At your next party try brontophobia (fear of thunderstorms), emetophobia (fear of vomit), trypanophobia (fear of injections), and triskadekophobia (fear of the number thirteen).

POST-TRAUMATIC STRESS DISORDER

THE BASICS

Remember when we talked about how emotional events form much stronger memories than nonemotional ones? Post-traumatic stress disorder (PTSD) is an extreme example of that effect. PTSD is a response to a serious, often life-threatening, shock.

A traumatic event causes the body's fear and arousal systems to jump into action. PTSD seems to result from a failure to shut these systems off after the threat has passed. For people with PTSD, memories of the ordeal remain vivid and immediate, suffering uncontrollable flashbacks or nightmares. (Some people report that they feel as though they are actually reliving the ordeal.) Many with PTSD complain of feeling emotionally numb and disinterested in the world around them. They may also be in a near-constant state of arousal, being unable to sleep, anxious, and easily startled.

Of course, not everyone who experiences trauma goes on to develop PTSD (about 25 percent do). Whether someone develops PTSD is probably the result of the complex interplay of a variety of factors, including the nature of the trauma, individual psychiatric history, previous experiences with stress, and any genetic predispositions to anxiety or depression. Treatment usually includes anti-anxiety medications combined with talk therapy.

One of the most interesting new ideas for treating PTSD involves administering drugs known as beta-blockers, which are commonly used to treat high blood pressure. Beta-blockers also interfere with neurotransmitters that aid in memory formation, and studies have shown that the drugs can indeed disrupt the process by which memories are stored. Researchers have become interested in the possibility that beta-blockers could help people avoid forming strong memories of a traumatic event.

Clinical studies of this idea have yielded mixed results so far, but scientists are pressing on.

COCKTAIL PARTY TIDBITS

❖ The National Institute of Mental Health reports that nearly 20 percent of Vietnam War vets developed PTSD.

❖ The terrorist attacks on the United States on September 11, 2001, also caused PTSD nationwide. Psychologists did extensive studies showing that 11 percent of New Yorkers went on to develop PTSD in response to the attacks, compared to just 4 percent of the general U.S. population. Those who watched a lot of television coverage of the events of that day, as well as people who had friends and relatives die in the attacks, were at particularly high risk.

AMNESIA

The initials *H.M.* are familiar to every student of neuroscience—they represent perhaps the most famous patient in the field. H.M. suffered from severe epilepsy, and in 1953 doctors trying to control the seizures removed part of his brain, including his hippocampus, amygdala, and other areas involved in memory formation and storage.

Ever since then, H.M. has been stricken with profound amnesia. He has both retrograde amnesia, which means he can't recall certain events from his past, and anterograde amnesia, which means he can't create any new memories.

Studies of H.M. and other amnesiacs have revealed much about the way the brain works. (And, of course, provided tidbits for thousands of hours of soap operas.) One of the most important findings was that, although H.M. could not for the life of him learn and remember new facts or events, he could be taught new physical skills. He could learn to juggle without any memory of having learned how. Findings like these have helped scientists determine that the brain has different circuits for storing declarative memories (places, names, dates), and nondeclarative memories (how to ride a bike).

Though H.M.'s affliction was caused by physical damage to the brain, amnesia can also be caused by temporary brain injuries, such as concussions or psychologically traumatic events. That's basically all there is to say about amnesia. Unless I'm forgetting something.

ON THE FRONTIER

Scientists are still debating the mechanism that underlies amnesia. Is a patient who can't form new memories unable to create records of new events, unable to store these records, or merely unable to access them?

Many researchers maintain that amnesia is a problem with memory retrieval, pointing out that amnesiacs are sometimes able to recover "lost" information. But the issue is far from settled.

COCKTAIL PARTY TIDBITS

❖ Reverend Ansel Bourne was a pastor who disappeared from his Rhode Island home in January 1857. He had forgotten his name, identity, and all his prior memories—a condition known as dissociative fugue. For several months, he lived in Pennsylvania as a shopkeeper named A. Brown before suddenly remembering he was Ansel Bourne and wondering what he was doing in the Keystone State.

SCHIZOPHRENIA

Schizophrenia has often been portrayed with a fair bit of artistic license, but the media does get some things right: It's a disease marked by psychosis, or a considerable break from reality. Many people with schizophrenia have hallucinations, particularly auditory ones, such as hearing voices telling them what to do. They may also have delusions—some of the most common include delusions of being followed or persecuted as well as delusions that their thoughts are being controlled. Schizophrenics may also speak in a rush of tangential or incoherent thoughts.

But some schizophrenics don't have any of these symptoms. In fact, they may have symptoms that appear to be the exact opposite. They may withdraw entirely from the world, express little to no emotion, and perhaps not speak at all. They may also suffer from what is known as catatonic rigidity, holding the exact same, often awkward, pose for hours. This is a far cry from the homeless schizophrenics who wander the streets in episodes of *Law and Order*. Nevertheless, schizophrenia can profoundly derail a person's life, and the disease can be hard to treat.

Schizophrenia is a separate disease entirely from Multiple Personality Disorder (MPD), although the two are often conflated.

Schizophrenics have recognizable abnormalities in brain chemistry and structure, and many scientists have come to believe that some of these changes start in the womb. It is thought that exposure to viral infections may increase a fetus's chance of going on to develop schizophrenia.

New research even suggests that mothers who are exposed to severe stress—specifically, the loss of a loved one—during the first trimester of their pregnancies are more likely to have children who develop the disorder. Of course, not nearly all the children who are exposed to infections or stress in utero will be schizophrenic. But prenatal life is becoming a time of particular interest to researchers trying to pin down the causes of the disease.

COCKTAIL PARTY TIDBITS

❖ Babies born in spring and winter are more likely to develop schizophrenia.

❖ Schizophrenia becomes increasingly common as socioeconomic class decreases. It is entirely possible, however, that the disease leads people to make less money, rather than the other way around.

❖ Auditory hallucinations are actually marked by an increase in activity in the brain areas that process real sounds.

ADDICTION

Addiction is a dependence on drugs or another chemical substance, a dependence that can threaten health and disrupt daily life. Though different drugs work in different ways, at the most basic level, drugs feel good (Shhh, don't tell the kids!) because they activate the brain's reward center.

In general, drugs produce their rewarding effects by influencing the activity of certain neurotransmitters in the brain. Over time, the brain will compensate for the drug by changing its production of these neurotransmitters. If a drug consistently increases the level of dopamine in the synapses, the brain will adjust by making less of it. This gives rise to the phenomenon known as tolerance, where an addict needs more and more of a drug to feel the same effects. (It can also explain the awful feelings of withdrawal. If the drug suddenly disappears, the brain has to adjust back to its baseline level of activity.)

Addiction can and does destroy lives, and addicts may spend all their money on drugs, lose their jobs, and let their personal relationships collapse. Though addiction can strike anyone, some people seem to inherit genetic predispositions to become dependent on all sorts of substances. There are many treatment options for addiction (like the famous twelve-step programs), but it remains an insidious and difficult problem to beat despite the best efforts of modern medicine, neuroscience, and psychology.

ON THE FRONTIER

Scientists are still looking for a wonder drug to help addicts kick the habit. One promising approach involves preventing drugs from producing their pleasurable effects. Recent research in this area focused on the ventral tegmental area, which is part of the brain's reward pathway.

Scientists have discovered a naturally occurring enzyme that seems to boost activity in the ventral tegmental area. They then designed a substance that blocks this enzyme in the brain. Giving this substance to rats eliminated drug cravings and withdrawal symptoms.

COCKTAIL PARTY TIDBITS

❖ Addicts often strongly associate environmental cues with their drugs of choice. Just seeing a photo of the place where he normally takes drugs can stimulate an addict's cravings.

❖ In 1995, talk-show queen Oprah Winfrey revealed that she had used crack cocaine when she was in her twenties. She said she had started using the drug because of a lover and that she was "addicted to the man."

PATHOLOGICAL AGGRESSION

Aggression is a behavior that evolution kept around for a reason. But in modern human society, it's not always as adaptive as it once was. (Today it's more likely to get you a stint in jail than a piece of new territory.) A small percentage of the population commits the vast majority of all violent acts, and neuroscientists are trying to determine what's so different about the minds of this minority.

Unsurprisingly, the amygdala—which is involved in fear, anger, and the fight-or-flight response—is an important brain structure for violent and aggressive behavior. People who commit impulsive acts of violence have higher rates of glucose metabolism in their amygdalas than others, meaning that it's burning through fuel more quickly.

The frontal cortex usually helps us exercise self-control, and scientists believe that it might act as a brake on aggression, keeping our impulses in check. Research has revealed that this region can be underdeveloped or underactive in violent criminals. The neurotransmitter serotonin also has a role, and low levels of the substance have been found in aggressive people (as have mutations in genes that influence serotonin production).

A number of environmental risk factors might play a part as well. Poverty, childhood abuse, family dysfunction, drug use, and witnessing violent acts all increase a child's likelihood of later becoming violent himself. But even taking all the biological and environmental factors into account still doesn't allow us to predict who will become a violent offender.

A recent study of animals examines the differences between functional and dysfunctional aggression. Scientists allowed rodents to repeatedly dominate other rats and mice. This pattern of domination turned normal rodents into pathologically violent creatures.

Researchers also found that the animals that became pathologically violent had a precipitous drop in their serotonin levels. But healthier, more functional acts of aggression were not associated with this serotonin deficiency.

COCKTAIL PARTY TIDBITS

❖ Testosterone seems to play a role in pathological aggression. A study of prison inmates revealed that men who had committed the most violent offenses (or just violent offenses in general) had the highest levels of testosterone.

❖ People who are antisocial and prone to criminal activity tend to have low resting heart rates. This might represent a certain fearlessness.

SLEEP DISORDERS

Some people have long-term, even lifelong, difficulties falling or staying asleep. The neurological basis of chronic insomnia is not yet clear, but it probably results from abnormalities in the brain networks involved in arousal and alertness. There's much talk these days about "sleep hygiene," and insomniacs are instructed to try changing their sleep behaviors—by going to bed at a regular time, avoiding alcohol and caffeine in the evenings, and the like—before trying sleeping pills.

Narcolepsy is a sleep disorder of a different ilk, characterized by severe sleepiness during the day, so much so that narcoleptics are often forced to take several daily naps. The other notable symptom of narcolepsy is sudden muscle weakness, in response to strong emotions. This phenomenon, known as cataplexy, might force a narcoleptic to feel weak at the knees or even collapse during times of fear, laughter, or embarrassment. Narcoleptics have a deficiency of a hormone known as orexin, which is made by cells in the hypothalamus, which helps regulate sleeping and waking.

The gold standard for diagnosing the whole gamut of sleep disorders is polysomnography, which involves an overnight stay at a sleep clinic. Patients are hooked up by electrodes to machines that monitor their sleep stages in the hopes that doctors will be able to determine what their brains are up to as they sleep, or simply try to.

Researchers believe that by artificially depressing orexin activity, they might be able to help insomniacs feel sleepy.

Adenosine is also known to be involved in sleepiness. A second possibility might be to treat insomnia by injecting adenosine into the brain or boosting its action. Enter, Sandman.

COCKTAIL PARTY TIDBITS

❖ More than 20 percent of the population suffers from insomnia in any given year. Considerably fewer people—about 4 to 5 percent—say they feel overly sleepy during the day.

❖ In sleep paralysis, a sleeper is unable to move his body upon awakening. Scientists believe this occurs when the brain temporarily fails to switch off the physical paralysis it activates during sleep. The sufferer of sleep paralysis may also experience hair-raising hallucinations. Sleep paralysis might explain stories of UFO abductions and ghostly encounters.

PRION DISEASES

They're suffering from some of the strangest symptoms on the books: sheep that scratch off all their wool, cows that stumble and fall, humans with months-long insomnia that causes death. What unites them might be stranger still: malformed, infectious proteins known as prions (a combination of the words *proteinaceous* and *infectious*).

It was once accepted that the only things that could transmit disease were bacteria, viruses, parasites, and so on—things that had DNA. Prions don't. That's why nearly everyone rejected the idea that proteins could make people sick.

But prions, most scientists now admit, are the cause of a variety of fatal, degenerative neurological disorders. Proteins fold into complex and unique shapes. Prion diseases are caused when a specific protein takes on an abnormal shape, remaining unfolded. When this prion touches other, healthy proteins, it triggers them to unfold. These unfolded prion proteins accumulate in the brain, causing cell death. As patches of neurons die, the brain becomes riddled with holes and comes to resemble a sponge.

None of the human prion diseases are common, but the best known is Creutzfeldt–Jakob disease, which you probably know as mad cow disease. Humans can get it by ingesting prions present in infected cattle. The condition can take months or years to develop, but is untreatable and fatal. Prion diseases can also result from genetic mutations; fatal familial insomnia is one such disease.

Mad cow disease caused a big scare when people in the United Kingdom started showing symptoms of the human form of the disorder. The panic was so widespread that the Red Cross banned anyone from giving blood who had spent more than three months in the United Kingdom between 1980 and 1996, out of a fear that they might transmit prions through their blood.

Scientists have now created promising new technologies that can filter prions from blood. The breakthrough may greatly expand the pool of potential blood donors.

COCKTAIL PARTY TIDBITS

❖ Prions are hardy little suckers that can't be killed by sterilization, heat, or radiation.

❖ One of the most exotic prion diseases is Kuru, found in a native tribe of Papua New Guinea. The prions that cause Kuru mostly affect the cerebellum, causing trembling and motor impairment. It is believed that the disease spread so rapidly because the tribe ate the remains of their dead relatives, including the prion-loaded brains.

MULTIPLE SCLEROSIS

My third-grade teacher tried to write legibly on the blackboard, but the lessons were often hard to decipher. One of the parents discovered that she had multiple sclerosis, a degenerative disease that compromises movement, making it difficult for her to control fine motor skills like writing. Over the next few years, her handwriting got worse, and she deteriorated until she needed a wheelchair.

Multiple sclerosis (MS) is classified as an autoimmune disease, a disorder in which the immune system mistakenly attacks the body. In the case of MS, the immune system targets myelin, the fatty substance that insulates neurons to speed nerve signals. The deterioration makes nerve signals unreliable, and movements become impaired.

Patients with MS report muscle weakness and numbness, decreasing motor control, impaired vision, jerky tremors, and difficulty walking. In many people, attacks wax and wane with long periods of remission, but the condition is chronic. Steroids can help treat the attacks and reduce symptoms. MS has a genetic component, but genes alone aren't sufficient to cause the disease. The environmental triggers are still unknown, but some have suggested that exposure to certain viruses might contribute.

Men, especially those at their reproductive peak, rarely develop MS. Animal studies have shown that testosterone can protect against certain autoimmune disorders, and researchers have theorized that the sex hormone might be similarly effective in staving off MS.

Now, a small study has shown that testosterone can help alleviate its symptoms. Men with MS rubbed a testosterone gel on their arms every day for a year. The men had increased muscle mass, reduced symptoms, and even slowed rates of brain deterioration.

COCKTAIL PARTY TIDBITS

❖ The incidence of MS increases with latitude—the disease becomes more common closer to the world's poles. (Even within the United States, the disease is far more prevalent in Minnesota or North Dakota than in Florida or Alabama.) Scientists aren't sure why this is, but this well-documented phenomenon has been cited as evidence that environmental factors must play a role in the disease.

❖ Early autopsies of patients with MS revealed that areas of their brains were hard and covered with scar tissue. This observation gave rise to the name multiple sclerosis (*sclerosis* means *scar* in Latin).

PARKINSON'S DISEASE

They come from different worlds, but celebrities Muhammad Ali and Michael J. Fox have been joined by their efforts to put real, live faces on Parkinson's disease. Slowed movement, deteriorating posture, a masklike face, rigid muscles, and tremors are all hallmarks of the condition.

Compared to many other neurological disorders, the molecular basis of Parkinson's disease is well understood. Patients have extensive cell death in part of the brain involved in motor control, the basal ganglia. When they're healthy, these neurons release dopamine to communicate with parts of the brain that direct bodily movements.

When these neurons die, dopamine levels fall. Walking, getting dressed, and other behaviors become impaired. Though Parkinson's is incurable, doctors do have treatments at their disposal for many of the symptoms such as L-DOPA, which the brain converts into dopamine. But L-DOPA isn't a permanent solution and has serious side effects, so scientists are trying to develop other drugs that mimic the action of dopamine.

For patients with particularly severe symptoms, surgeons can destroy a small group of cells that inhibits movement. Also, a procedure known as deep brain involves an electrode implanted in the brain to help stimulate certain parts. More on that in the next chapter.

More on that in the next chapter.

ON THE FRONTIER

Several studies have revealed that over-the-counter medications might have a variety of uses in combating Parkinson's and its symptoms. Pain relievers like Advil, known as non-steroidal anti-inflammatory drugs, or NSAIDS, seem to reduce a person's risk of developing the disorder.

And cough medicine—that gross, cherry-flavored stuff we've all choked down—might help suppress what are known as dyskinesias, serious motor side effects of L-DOPA. (Dyskenesias are involuntary movements and can be so disabling that they actually keep people from taking an otherwise effective medication.)

COCKTAIL PARTY TIDBITS

- ❖ Parkinson's and L-DOPA were the subject of Oliver Sacks's book *Awakenings*, and the subsequent movie starring Robin Williams.
- ❖ In 1982, a handful of young drug addicts in California mysteriously developed a syndrome resembling Parkinson's. Medical detective work eventually uncovered that the patients had bought synthetic heroin on the streets contaminated with a substance known as MPTP. MPTP induces Parkinson's-like symptoms.

ALZHEIMER'S DISEASE

On November 5, 1994, Ronald Reagan released a handwritten letter to the American public. In it, the former president confessed to having one of the most devastating neurological diseases: Alzheimer's. A degenerative disease, Alzheimer's is estimated to afflict four or five million Americans.

The earliest symptoms—forgetfulness, disorientation, and difficulties finding the right word—are mild and sometimes hard to distinguish from normal signs of aging. But as the disease progresses, its distinct characteristics become apparent. Memory deteriorates, as does judgment and other thought processes. In the end stages, the disease is so severe that Alzheimer's patients may not recognize their own family members or be able to take care of themselves.

Studies of the brains of Alzheimer's patients have revealed several characteristic problems. They have cell death, deposits of proteins around the synapses called amyloid plaques, and protein tangles inside the neurons called fibrillary tangles. Scientists believe these plaques and tangles cause neurons to die, but they're still exploring the exact role they play.

People with Alzheimer's also have reduced levels of certain neurotransmitters, including one involved in learning and memory called acetylcholine. The cause of Alzheimer's is still a mystery—there are many theories—but some people appear to have a genetic predisposition for an early onset form. There are no cures or effective treatments yet.

ON THE FRONTIER

The only way to conclusively determine whether a patient has Alzheimer's is to examine the patient's brain after death, looking for buildup of plaques and tangles. But researchers have recently made progress in developing tests that can diagnose the disease while the patient is still alive. After all, a treatment will do us no good if we can't figure out who to give it to.

A chemical tracer used in brain imaging actually binds to amyloid protein, which is present in plaques. That means doctors can inject the substance into patients' bloodstreams and then use PET scans to see whether it turns up in the brain. If it does, it means plaques could be present and the patient may be on her way to developing full-blown Alzheimer's.

COCKTAIL PARTY TIDBITS

❖ Up to 80 percent of dementia cases in the elderly can be attributed to Alzheimer's.

❖ Women with Alzheimer's tend to survive longer than men with the disease.

❖ Alzheimer's causes a decline in patients' ability to smell.

STROKE

Strokes are caused by a sudden lack of blood—and thus oxygen—to the brain, which kills neurons. Usually, a blood clot blocks the flow of blood, but strokes are also caused by bleeding in the brain.

Signs of a stroke include numbness on one side of the body, a splitting headache, trouble speaking or understanding speech, blurred vision, and dizziness, depending on where in the brain the blockage occurs. High blood pressure and cholesterol, smoking, diabetes, and obesity can all increase a person's risk, and genetically abnormal blood vessels can also make strokes more likely.

Strokes can cause long-lasting disability if they are not treated quickly. Doctors use brain imaging to reveal the type and location of a stroke, and then select a treatment based on what caused the stroke in the first place. Surgery can be used to clear out a clot or reopen a blocked artery. Doctors may also administer medication that keeps the blood from clotting to prevent future strokes.

Aspirin should never be taken at the first sign of a stroke, because if the blood loss is caused by hemorrhaging instead of a clot, it can make the condition worse. If the stroke is due to bleeding, doctors may have to drain out the blood.

Even with these interventions, some people who have

strokes have lingering difficulty performing tasks such as speaking, getting dressed, or walking. Physical therapy can sometimes help patients regain these skills.

Doctors and engineers have teamed up to create a computerized system they hope will allow stroke victims to rehabilitate faster. The program is designed to improve motor skills by having patients perform tasks in virtual reality.

One component, for instance, involves using a joystick to manipulate objects in the virtual environment. Scientists believe that doing so will improve patients' hand–eye coordination, which can be damaged by a stroke.

COCKTAIL PARTY TIDBITS

❖ Stroke is the third-most common cause of death in the United States. Some 750,000 people have strokes each year.

❖ Some people who have strokes in the parietal lobe suffer from a condition known as hemispatial neglect, which causes them to ignore an entire half of their visual fields. For instance, if someone has a stroke that damages the left side of the parietal lobe, they might not be able to "see" things on their right side. They might only eat food on the left side of their plates and only shave the right side of their faces (on the left in the mirror).

HEADACHES AND MIGRAINES

Most of us have a lifetime of familiarity with headaches, which have a tendency to make an appearance at the most inappropriate times. Brain tumors can bring on headaches, but don't assume the worst just because your head is pounding—headaches can also be triggered by stress, hunger, and even sex.

There are many different kinds of headaches. Tension headaches result from tightening muscles in the face and neck; many other types of headaches are caused by the swelling of blood vessels in the head. (Caffeine, which causes blood vessels to contract, often helps these headaches.)

There are headaches, and then there are *headaches*. Migraines are freight-train-in-your-brain, type-it-in-italics *headaches*, characterized by severe (sometimes incapacitating) pain and often accompanied by nausea and sensitivity to light and noise. Many migraine sufferers experience auras—visual hallucinations like shimmering or flashing lights—shortly before a headache comes on.

Certain foods—including onions, bananas, and cheese—are known to trigger migraines, and sufferers may be able to alleviate their symptoms by eliminating the right dietary offenders. Also, women are more likely to suffer from migraines, which seem especially prevalent around menstruation. Researchers are still exploring the links between migraines and hormones.

ON THE FRONTIER

Teenagers who suffer from chronic migraines are more likely to commit suicide, a recent study shows. And it's not just because of the crushing headaches. Those who have the headaches are also at increased risk for mood disorders, panic disorder, and social phobia. These connections are well documented, but scientists are unsure how to explain them. Many of these disorders are associated with abnormal levels of serotonin, which could explain the links.

COCKTAIL PARTY TIDBITS

- ❖ Want to use your headache to get out of doing something? Call it by its medical term: cephalgia. It sounds much more serious that way. "I can't come to work today because I have cephalgia."
- ❖ Accountants get more headaches than people in other occupations, according to a survey. And who can blame them? Librarians, bus drivers, and construction workers were similarly afflicted.
- ❖ Hangovers result from the swelling and irritation of the brain and its blood vessels. Wine is often blamed for the worst hangovers, but no matter what your poison, your best bet is to drink responsibly and stay well hydrated.

HEAD INJURIES

The brain is usually protected by the skull, but a sudden or violent force can turn the skull from a shield into a battering ram. This collision can bruise the brain or cause it to swell, tear, or bleed. (My head hurts just thinking about it.)

Concussions are the least serious of these traumatic brain injuries and can be caused by something as innocent as childhood sledding (and I'm speaking from experience here). Concussions usually cause a temporary loss of consciousness, and can also result in amnesia, nausea, and headaches (but not, alas, stars or birds revolving around your head). They are often mild, and recovery can seem complete. But the brain damage caused by concussions is cumulative, which means that football players who step back onto the field after a concussion are at increased risk for serious repercussions the next time they suffer a head injury.

Although concussions are closed head wounds, the brain can also be injured by an object piercing the skull. One of the classic causes of these penetrating brain injuries is a gunshot wound to the head, which can cause a bullet to become lodged in the brain's tissue. Believe it or not, doctors might operate on such victims to clean the wound, but they'll often leave the bullet in place; removing it can cause more brain damage than just leaving it be. (That's a pretty cool souvenir, though not necessarily one that's easy to get through airport security.)

Bad news, professional football players (are any of you reading this?): The concussions they accumulate during their athletic careers may lead to dementia. A study of more than twenty-five hundred retired NFL players revealed that those who had sustained at least three concussions were five times more likely to be diagnosed with cognitive impairments and three times more likely to suffer from memory problems as they aged. And retired football players as an entire group are 37 percent more likely to develop Alzheimer's.

COCKTAIL PARTY TIDBITS

* Men are at least three times more likely to suffer from traumatic brain injuries than women. I guess playing football and riding motorcycles will do that.
* We were almost born into a world without football. In the early twentieth century, President Teddy Roosevelt threatened to outlaw the game because players, who didn't wear helmets at that time, were accruing serious—even fatal—brain injuries. As many as two dozen players could die in a single season from such trauma. That sort of makes today's players sound like total wimps.

COMA

Serious brain injuries can cause someone to lapse into a long period of unconsciousness, or coma. People in comas are unresponsive and seem unaware of their surroundings. Their bodies may still be capable of basic functions, but their brains are essentially hibernating.

Some people in comas eventually fall into a "persistent vegetative state." In this condition, patients might open their eyes but will generally still fail to respond to the world around them. That doesn't mean that they remain still and silent, however, and they may make unprompted movements—move their eyes or limbs, grunt, and so on.

The prognosis for people in comas or persistent vegetative states varies greatly. In general, the longer someone spends in one of these states, the less likely a meaningful recovery becomes. Some people do awaken from comas and vegetative states, but others spend years hooked up to ventilators with little chance of improvement. Some people choose to draw up living wills that spell out their wishes should they be unfortunate enough to end up in such condition.

Distinguishing between several comalike states is difficult for doctors; predicting which patients will recover (and how well) is tougher still. But researchers have recently developed a new tool for assessing patients in comas and making solid forecasts about the extent of their recovery.

In a study, scientists showed that their "Disorders of Consciousness Scale"—which involves assessing a patient's sensory systems, ability to swallow, balance, and more—was able to reliably predict which patients would improve within a year of their injuries.

COCKTAIL PARTY TIDBITS

- That study of comas in soap operas? It found that the portrayal of such states of unconsciousness was merely coma ci, coma $ça$. (Sorry—couldn't resist.) In short, these portrayals were too optimistic, failing to show some of the lingering difficulties experienced by people who do awaken from comas.

- In 2003, an Arkansas man who had spent nineteen years in a comalike state (yes, that's *nineteen years*) spontaneously awoke. The man woke with amnesia, and was shocked to hear that Ronald Reagan was no longer the president. Him and the entire GOP.

IT'S (USUALLY) NOT BRAIN SURGERY: MENDING THE MIND

PSYCHOTHERAPY

Psychotherapy is a classic intervention for all sorts of mental disorders. It generally involves repeated, one-on-one sessions between a patient and a therapist. These sessions feature the old-fashioned art of talking: patients are encouraged to discuss their lives, problems, hopes, dreams, and so on. But beyond that there's a therapeutic style and philosophy to suit everyone.

For instance, there's Freudian analysis, where the therapist analyzes your free associations, dreams, and unconscious beliefs. At the other end of the spectrum is cognitive behavioral therapy (CBT), which wastes little time on a patient's past or his unconscious desires. Rather, CBT attempts to rewrite a patient's destructive thought patterns and actions.

Therapists are bound by strict ethical codes that compel them to keep sessions confidential and forbid them from doing inappropriate things like seducing their emotionally vulnerable patients. Psychotherapy is exceedingly expensive, and even patients who are insured often don't have their sessions fully covered. And psychotherapy may be utterly out of reach for patients with no insurance at all. Group therapy, in which many patients work together with one shrink, is a more affordable option for some.

One study found that monthly "maintenance" sessions of psychotherapy could help prevent relapses among women with recurrent depression. Monthly appointments may be a good alternative (or even supplement) to recommendations that patients with recurrent depression stay on antidepressants for long intervals.

Another study found that depressed teens who combine antidepressant medications with psychotherapy do much better than those who rely on either drugs or therapy alone. Even therapy administered over the telephone was shown to help.

COCKTAIL PARTY TIDBITS

❖ Even mental health is being outsourced these days, as some patients choose to participate in cybertherapy, mental health treatment through therapists with online practices. But many therapists that continue to treat patients exclusively in the real world have expressed concerns about the quality of care delivered by these e-therapists, who charge as much as two dollars a minute.

❖ Groups have sprouted up to provide therapy for nearly every conceivable problem. One of the latest: group therapy for men suffering from erectile dysfunction. Studies even reveal that it helps.

PSYCHOPHARMACOLOGY

Doctors and psychiatrists are assuredly prescribing mind-altering drugs to patients who don't need them. But they are also prescribing them to many patients who do, and psychiatric medications indisputably save lives on a regular basis. There are medicine cabinets full of psychiatric drugs, and there's no one-size-fits-all solution. The pill that works great for Joe may be terrible for Karen, even though they're both paranoid schizophrenics. And there are few, if any, magic bullets.

Though the meds can bring great relief, they often come with a hefty dose of side effects, which, depending on the pill, can range from tremors to sexual dysfunction. Even if a medication works wonders, doctors often face challenges of compliance—patients who start to feel better may abruptly quit their meds, triggering a serious relapse.

A major obstacle to the development of new psychiatric drugs is the blood–brain barrier, a semipermeable membrane that blocks many substances in the blood from entering the brain. This barrier usually works to protect the brain. But it can thwart the development of new therapeutic agents, which have to be able to pass through this membrane in order to even have a shot at being effective.

Nanotechnology, which involves artificial devices that

operate on the molecular level, could be of great use in this area; researchers hope it will allow them to develop tiny particles that can cross the blood–brain barrier and deliver drugs.

ON THE FRONTIER

One of the major controversies over psychiatric drugs in recent years has been the apparent link between antidepressants and suicide in children and teenagers. In more than one high-profile case, a young person committed suicide shortly after starting antidepressants.

It's thought that the risk of suicide may increase as depressed patients start to come out of their stupor and their energy levels increase. In fact, many of the most severely depressed patients may not have the energy to plan or carry out a suicide attempt. The FDA now requires that all antidepressants carry a so-called black box warning cautioning that the medication could lead to suicide attempts.

COCKTAIL PARTY TIDBITS

- One in ten American women take antidepressants. One in twenty American men do.
- Lexapro, a medication for depression and anxiety, was the sixth most frequently prescribed psychiatric medication in the United States in 2006. Twenty-six million prescriptions for it were written that year.

BRAIN SURGERY

Despite the many breakthroughs of modern neurosurgery, lobotomy remains the discipline's most infamous procedure. And with good reason: There were a variety of different techniques, but perhaps the best known involved inserting an ice pick through the eye socket. The end of the pick was then moved around in the brain to destroy tissue and nerve fibers.

The procedure was devised to treat all manner of mental illnesses but caused problems worse than the ones it was designed to fix. Lobotomized patients could suffer from paralysis, cognitive and emotional impairments, seizures, and more. The surgery caused such horrible "side effects" (if you can call them that) that it eventually fell out of use.

Fortunately, we now have other surgical techniques for mending the mind. For obvious reasons, they are not the first line of defense for many disorders, but they can help when other interventions fail.

Most types of brain surgery involve literally removing the problem—a tumor or a bit of malfunctioning tissue. But surgeons have to be exceedingly careful to remove only what is necessary, which is why brain surgery is usually performed with the patient awake. For instance, if a surgeon is trying to excise a tumor that resides near one of the brain's language-processing centers, he might ask the patient to perform language tasks during the operation. If the patient is suddenly unable to do so, the surgeon will know she's gotten too close to a critical area. Surgeons also use sophisticated brain imaging during surgery to help them pinpoint the exact locations for their scalpels. We've come a long way since the ice pick.

ON THE FRONTIER

One of the most cutting-edge techniques in brain surgery involves the use of a surgical instrument known as an endoscope. Endoscopy allows surgeons to perform minimally invasive procedures without gaping holes.

This technique is commonly used in medical procedures performed on the esophagus, stomach, and colon; now it has been used on a small number of children with brain tumors. The technique allows surgeons to remove brain tumors as large as baseballs through the nose!

COCKTAIL PARTY TIDBITS

❖ Operations on the brain can take eight or even twelve hours.
❖ The last thing you want to do while undergoing brain surgery is suddenly jerk your head to the side. That's why patients often have their heads screwed into a metal frame. This involves drilling holes into the skull. It doesn't cause any pain, thanks to a local anesthetic, but it makes noise and produces a burning smell.

ELECTROCONVULSIVE THERAPY

Eectroconvulsive therapy (ECT) sounds barbaric. (For a long time, it was colloquially called "shock therapy," after all.) Electrodes are attached to your head. You get muscle relaxants so you don't convulse and injure yourself and general anesthesia because otherwise what's about to happen would scare the living daylights out of you.

After you slip into unconsciousness, the electrodes are turned on, sending an electric current zapping into your skull. The current will cause neurons all over your brain to fire off their own electric signals and release all sorts of neurotransmitters. In response, you'll have a seizure. After the treatment, you'll wake up disoriented and confused and will likely have some short-term memory loss. Multiply that by the 6 to 12 treatments you'll need.

But here's the thing about ECT: It works. (Scientists still aren't sure why, but it does.) Somewhere in the neighborhood of two-thirds of patients who undergo the procedure find that their depression lifts afterward. And many of these patients are those who found absolutely no relief from other treatments. The technique has also been improved greatly in recent years. (It really used to *be* barbaric—now it just sounds that way.) ECT is still burdened by a not-quite-sterling reputation, but it remains a powerful intervention of last resort for patients suffering from severe depression.

Though ECT has been used primarily to treat depression, it might also help the symptoms and daily functioning of schizophrenics. Though antipsychotic medications are still the first line of defense for schizophrenia, combining them with electroconvulsive shock therapy seems to be even more effective, particularly in spurring rapid improvement. Twenty percent of schizophrenics do not respond to medication, and adding a round (or several) of ECT could help them find relief.

COCKTAIL PARTY TIDBITS

- ❖ In the early twentieth century, doctors became interested in the idea that seizures could help treat mental illness. They were exploring the idea of using chemicals to make the brain seize until Italian psychiatrist Ugo Cerletti had the idea that electric currents could do the same thing when he visited a slaughterhouse where pigs were having their heads shocked—and surviving. That was all Cerletti needed to convince himself that a similar approach might be feasible on humans.

- ❖ Before doctors thought to administer muscle relaxants, ECT used to cause patients to convulse so hard they would dislocate shoulders and break bones. That can't have helped anyone's depression.

DEEP BRAIN SIMULATION

For decades, we've had pacemakers for the heart—now we have them for the brain. The technology is called deep brain stimulation: A surgeon implants an electrode in the site of abnormal activity. This electrode is then connected by a wire to a power pack embedded under the collarbone.

When the power pack is turned on, it sends electrical stimulation to the problem spot. The pack has tens of thousands of possible settings, controlled by the patients themselves (under their doctor's supervision, of course).

Deep brain stimulation has proven effective in treating several disorders. It's been used most extensively to target the symptoms of Parkinson's disease. Electrodes are placed in one of the brain structures involved in movement. When the pacemaker is turned on, it blocks some of the abnormal neuron signals, alleviating tremor, stiffness, walking problems, and more.

Deep brain stimulation has several things going for it. First, the procedure is reversible. If the patient doesn't tolerate it well or wants to try a new therapy, the electrode and the rest of the parts can be removed without damaging the brain. Second, because the power pack has many different settings, it can be adjusted as the patient's condition changes.

Scientists are turning up tantalizing evidence that deep brain stimulation can rapidly alleviate depression. To treat depression, the electrodes are implanted deep in the brain, in a region called area 25—a part that seems to be abnormal in people who are depressed.

When the device is activated, the region is bombarded with signals, and patients who have spent years fighting against depression report that they can finally feel it lifting. Clinical trials have been small so far, and much work remains to be done before deep brain stimulation can become a mainstream treatment for depression. But scientists—and patients whose depression has not responded to conventional treatment—are hopeful.

COCKTAIL PARTY TIDBITS

❖ Deep brain stimulation may be of use in reducing the tics experienced by those with Tourette's syndrome. Researchers are also exploring using the technology to treat obsessive-compulsive disorder, chronic pain, and epilepsy.

❖ In 2007, researchers even used deep brain stimulation to treat a brain-injured man who was barely conscious. After the electrode was implanted, the patient was able to execute previously impossible tasks, such as chewing his food and naming objects in front of him.

TRANSCRANIAL MAGNETIC STIMULATION

One of the newest techniques of brain therapy is transcranial magnetic stimulation (TMS). Like electroconvulsive shock therapy, TMS causes electrical activation of the brain. However, the technique is substantially different (for which we can probably be grateful).

In TMS, wire coils placed above the head issue brief magnetic pulses that travel through tissue and bone and into the brain. These pulses induce neurons to produce small electrical signals. Researchers still don't know exactly why, but these signals seem to help the brain heal itself. So far, TMS has been tested as a treatment for a range of neurological and psychiatric conditions and appears to be particularly promising for depression.

Researchers have high hopes for TMS, which unlike ECT, allows the stimulation of one specific brain region, giving doctors precise control over the effects. Speaking of effects, TMS has few unpleasant ones: It is painless, does not require anesthesia, and does not cause memory loss.

Alas, TMS is not perfect. Though it can be used on any neural circuits in the cortex, the magnetic field does not extend very far into the brain. That means the technology can't be used to treat problems with structures located deep in the brain. And though many studies have shown that TMS is an encouraging therapy, it is still experimental.

ON THE FRONTIER

TMS is so experimental that pretty much everything about it is "on the frontier." But I suppose the use of the technology to enhance the brain function of healthy individuals is even more cutting-edge than the other applications. TMS can be used to make neurons fire repeatedly, strengthening some of the brain's connections.

The technology might be used to reshape brain circuits in ways that improve human performance on a variety of tasks. Indeed, some research has already shown that TMS of the prefrontal cortex can improve people's puzzle-solving skills. And the government is funding research into whether TMS might be able to help rejuvenate fatigued soldiers on the battlefield. Call it the new stop-loss initiative.

COCKTAIL PARTY TIDBITS

* TMS apparently feels like a light "tapping" on the head.
* Some researchers and entrepreneurs are now working to create a portable TMS device that could be mounted in a helmet so you can take that happy feeling with you wherever you go. If only it could make you indifferent to being mocked for wearing a helmet.

STEM CELLS

Stem cell therapy is one of the great hopes of modern neuroscience, and researchers are particularly interested in harnessing the potential of the stem cells that exist in embryos. Embryonic stem cells do not yet have specific jobs in the body—their role is to grow into all the diverse and specialized cell types that an embryo will need.

The fact that embryonic stem cells can grow into any kind of cell means that they could be used to replace damaged cells in adults. Indeed, researchers have already shown that they can harness this potential; scientists can harvest the cells, spur them to replicate in a lab, and then coax them into becoming anything from blood cells to neurons.

The promise of stem cell therapy is especially strong for the brain. Unlike cells in many other parts of the body, damaged or dying neurons generally aren't replaced, making brain and spinal cord injuries particularly devastating. But stem cells could help stack the odds in our favor.

So far, research has shown that the brain can indeed take up stem cells from transplants and incorporate them into its circuits. The hope is that researchers will be able to use stem cells as a source of replacement neurons in problems ranging from Parkinson's to hearing loss. Encouraging, but there are a lot of technical difficulties to overcome before stem cells come to a brain near you.

Because of the ethical brouhaha surrounding the collection of embryonic stem cells, some researchers are now researching the potential of doing transplants with stem cells harvested from adults. (Certain tissues in our body have stem cells waiting to be called into action when they are needed.)

But adult stem cells tend to turn into whatever kind of tissue they live in—blood stem cells turn into adult blood cells, for instance. Scientists are currently exploring the limits of these stem cells, with some success. They have already managed to turn blood stem cells into neurons. Future research will provide a better idea of the potential of this approach.

COCKTAIL PARTY TIDBITS

❖ As you are probably aware, the use of embryonic stem cells is controversial. Preferring not to leave the scientific decisions to those who actually, ahem, know the science, President George W. Bush signed an executive order banning federally funded researchers from creating new lines of embryonic stem cells. Look for this ban to be overturned in the near future.

GENE THERAPY

Despite the awe with which we discuss them, genes are just a set of instructions, similar in principle to the little booklet that comes with an IKEA table you have to assemble yourself. What do you do when you follow the instructions but keep on churning out wobbly three-legged tables? You could pay a carpenter to add a fourth leg to all your tables, buy a bunch of new tables that are already put together, or go to the source and get a correct set of instructions. The last approach is the idea behind gene therapy.

Some brain diseases are caused, at least in part, by errors in the genetic instructions that tell the body how to build proteins. An error can cause the body to produce deformed proteins, the wrong protein, or no protein at all, wreaking havoc on the brain. Gene therapy involves sending a corrected set of genetic instructions into the body. To do so, scientists insert a copy of a healthy gene into a vector, an agent such as a virus that is capable of carrying the gene into the body. Once there, the new gene can insert itself into the cells' DNA and start cranking out normal proteins.

Gene therapy is still experimental, but recent research shows its potential to battle brain diseases. Scientists have already successfully gotten genes of interest to travel to the brains of rodents and monkeys and start manufacturing proteins. And the technique has been used to successfully improve the brain function of animals. Researchers have also shown that even a single injection of the genes is enough to make a visible difference in protein production in the brain. Many scientists have high hopes for using the approach to deliver genes that help treat all sorts of genetic neurological conditions, from Parkinson's to blindness.

COCKTAIL PARTY TIDBITS

❖ Jesse Gelsinger is a name that strikes fear in the heart of gene therapy researchers. Gelsinger was an eighteen-year-old who enrolled in a clinical trial of gene therapy in 1999. Several days after he started to receive injections of new genes, he died. His death, which may have been the result of a drastic immune-system response to the vector used to carry the genes, put a major freeze on gene therapy research.

ROBOTIC LIMBS

One afternoon in 2000, an owl monkey named Belle moved her arm in Durham, North Carolina. It might not sound like much, but it was the arm movement heard round the world. At least, it was heard seven hundred miles away in Cambridge, Massachusetts, where a robotic arm moved in perfect synchrony with Belle's.

This was a dramatic demonstration of using technology to harness the directives of the mind. Since then, there have been more encouraging breakthroughs in the field of robotic limbs, or neuroprostheses.

We still have a long way to go before we start giving amputees robotic prostheses, but the work in animals has been promising. The technique, which seems borrowed directly from science fiction, requires researchers to implant an array of electrodes in the motor regions of the brains of animal.

A computer program records and analyzes the patterns of neuron firing that correspond to a movement of interest (reaching an arm out, grasping an object). After the computer has learned what neural signature to watch out for, scientists will hook it up to a robotic limb, located perhaps miles away from the animal.

When the computer recognizes the pattern of neuron activity that corresponds to an impending movement, it can signal the robot to begin the same one. This all happens fast enough that the robotic limb can actually move in real-time synchrony with the animal's.

Some of the coolest research into brain–machine interfaces is allowing people who are utterly paralyzed to type out their thoughts. Scientists outfitted a few such individuals with electrode caps that recorded their brain activity. To type, they used specially designed software that flashed individual characters one at a time. By concentrating on the letter they wanted to select, the participants generated a pattern of brain activity that signaled the computer to execute the proper keystroke. The technology is giving severely disabled people back the ability to use their personal computers—and to become as addicted to them as the rest of us are.

COCKTAIL PARTY TIDBITS

❖ Sometimes the animals don't even have to move a muscle in order to trigger a robotic arm to move. One study showed that monkeys could get a robotic arm to feed them fruit just by thinking about the movements involved. Next up: training a robotic arm to fan monkeys with palm fronds while they're enjoying their grapes.

THE POWER OF NURTURE: CHANGING BRAINS

NEUROPLASTICITY

Amputees sometimes experience phantom limb sensations, feeling pain, itching, or other impulses coming from limbs that no longer exist. Neuroscientist V. S. Ramachandran worked with patients with so-called phantom limbs, including Tom, a man who had lost one of his arms.

Ramachandran discovered that if he stroked Tom's face, Tom felt like his missing fingers were also being touched. Each part of the body is represented by a different region of the somatosensory cortex and, as it happens, the region for the hand is adjacent to the region for the face. The neuroscientist deduced that a remarkable change had taken place in Tom's somatosensory cortex.

Ramachandran concluded that, because Tom's cortex was no longer getting input from his missing hand, the region processing sensation from his face had slowly taken over the hand's territory. So touching Tom's face produced sensation in his nonexistent fingers.

This sort of rewiring is an example of neuroplasticity, the adult brain's ability to change and remold itself. Scientists are finding that the adult brain is far more malleable than they once thought. Our behaviors and environments can cause substantial rewiring of our brains or a reorganization of its functions and where they are located. Some believe that even our patterns of thinking alone are enough to re-shape the brain.

Scientists have developed a technique known as vision restoration therapy to help stroke patients with partial vision loss.

The therapy requires patients to look at hundreds of different images that are presented to them on a computer screen. The images are designed to engage the parts of the visual field that have been compromised and thereby stimulate the damaged neural circuits. Remarkably, just two thirty-minute sessions of such therapy each day help strengthen and repair these networks and at least partially restore vision.

COCKTAIL PARTY TIDBITS

❖ Even wasps (the insects, not the people on Martha's Vineyard) exhibit neuroplasticity. As wasps take on increasingly complicated jobs in their colonies, their brains actually get bigger.

❖ Some plucky entrepreneurs have exploited the idea of neuroplasticity to design video games to help our brains stay nimble. Most scientists, however, remain skeptical that these video games improve anything other than our skill at playing these video games.

NEUROGENESIS

If neuroscience ever had a dogma, it was this: We are gifted at the beginning of life with masses of neurons, but that's all we ever get. If we squander, mistreat, or destroy those neurons, too bad—we can't have new ones. But this dogma was wrong. The first compelling evidence for adult neurogenesis came from studies of birds—our feathered friends generate new brain cells as they learn to sing songs. Over the last two decades, researchers have demonstrated that neurogenesis happens in humans, too.

Researchers now know that neurogenesis is a normal feature of the adult brain. Studies have shown that one of the most active regions for neurogenesis is the hippocampus, a structure that is vitally important for learning and long-term memory.

Neurogenesis also takes place in the olfactory bulb, which is involved in processing smells. But not all the neurons that are born survive; in fact, the majority die. In order to make it, the new cells need nutrients and connections with other neurons that are already thriving. Scientists are currently identifying the factors that affect the rate of neurogenesis and the survival of new cells. Mental and physical exercise, for instance, both boost neuron survival.

Speaking of the survival of new neurons, research shows that stress can sabotage them. In a recent study, scientists arranged encounters between young rats and older, more aggressive rats. The older rodents often physically pinned down the younger ones and even bit them. (This, unsurprisingly, causes stress.)

The scientists found that the encounters didn't influence how many new nerve cells were born in the hippocampi of the youngsters, but it did reduce the number that survived. The researchers hope that understanding more about this mechanism could shed light on depression, a disease in which reduced neurogenesis seems to play a role.

COCKTAIL PARTY TIDBITS

- ❖ A number of studies of neurogenesis have been conducted on crayfish and lobsters. One found that reducing a lobster's level of serotonin reduced how many new nerve cells were born and survived in the crustacean. No word yet on whether neurogenesis makes lobsters more delicious.

- ❖ The birth of new neurons is influenced by the body's circadian rhythms. (We also know this from studying the lobster.)

EXERCISE

Mice that run on wheels increase the number of neurons in their hippocampi, and perform better on tests of learning and memory.

Studies of humans have shown that exercise can improve the brain's executive functions (planning, organization, multitasking, and more). Exercise is also well known for its mood-boosting effects, and people who exercise are less likely to get dementia as they age.

Among those who are already aged, athletic senior citizens have better executive function than those who are sedentary; even seniors who have spent their entire lives on the couch can improve these abilities just by starting to move more in their golden years.

A variety of mechanisms might be responsible for this brain boost. Exercise increases blood flow to the brain, which also increases the delivery of oxygen, fuel, and nutrients to those hard-working neurons. Research has shown that exercise can increase levels of a substance called brain-derived neurotrophic factor (BDNF), which encourages neuron growth, communication, and survival.

Of course, all this research does nothing to help explain dumb jocks. Or Richard Simmons.

New research shows a little music can make your workout better yet.

Volunteers completed two workout sessions. In one, they sweated to the sweet sound of silence; in the other, they listened to Vivaldi's *Four Seasons*. After each workout, participants completed assessments of their mood and verbal skills.

Exercise alone was enough to boost both, but verbal scores improved twice as much when the exercisers had tunes to listen to. Maybe you can get your insurance company to pay for a new iPod.

COCKTAIL PARTY TIDBITS

❖ Exercise also improves sleep quality, a whole pile of studies suggest. And immune function. Is there anything it can't do?

❖ You don't need to be Chuck Norris (thankfully) to get the brain benefits of exercise. Studies with senior citizens have shown that as little as twenty minutes of walking a day can do the trick.

DIET

The brain needs fuel just as the body does. So what will really boost your brain power and what will make you lose your mind? Saturated fat, that familiar culprit, is no better for the brain than it is for the body. Rats fed diets high in saturated fat underperformed on tests of learning and memory, while humans who live on such diets seem to be at increased risk for dementia.

Not all fat is bad news, however. The brain is mostly fat—all those cell membranes and myelin coverings require fatty acids—so it's important to eat certain fats, particularly omega-3 fats, which are found in fish, nuts, and seeds. Alzheimer's, depression, schizophrenia, and other disorders may be associated with lowered levels of omega-3 fatty acids.

Fruits and vegetables also appear to be brain superfoods. Produce is high in substances called antioxidants, which counteract atoms that can damage brain cells. High-antioxidant diets have been shown to keep learning and memory sharp in aging rats and even reduce the brain damage caused by strokes. That's food for thought.

It's not just what you eat that affects the brain. It's also how much. Research has shown that laboratory animals fed calorie-restricted diets—anywhere from 25 to 50 percent less than normal—live longer than other animals. And, it turns out, they also have improved brain function, performing better on tests of memory and coordination.

Rodents on calorie-restricted diets are also better able to resist the damage that accompanies Alzheimer's, Parkinson's, and Huntington's disease.

COCKTAIL PARTY TIDBITS

❖ Some of the best brain foods: walnuts, blueberries, and spinach.

❖ It's especially important that babies get enough fat. Babies who don't get enough of the stuff have trouble creating the fatty myelin insulation that helps neurons transmit signals. Luckily for babies, breast milk is 50 percent fat.

❖ Populations that traditionally eat diets high in omega-3 fatty acids (Inuit populations, for instance, which eat a lot of fish) tend to have lower rates of disorders of the central nervous system.

STIMULANTS

Stimulants are substances that rev up the nervous system, increasing heart rate, blood pressure, energy, breathing, and more. Caffeine is probably the most famous of the group. (It's actually the most widely used "drug," if you want to call it that, in the world.) By activating the central nervous system, caffeine boosts arousal and alertness. In high doses, though, this stimulation can go too far, causing jitters, anxiety, and insomnia.

Cocaine and amphetamines are less benign. Though they work on the brain through different mechanisms, they have similar effects. Taking them increases the release of some of the brain's feel-good neurotransmitters—including dopamine and serotonin—and produces a rush of euphoria. They also increase alertness and energy.

That all sounds pretty good, but in high doses they can cause psychosis and withdrawal. The withdrawal symptoms are nasty and can lead to depression, the opposite of that euphoric feeling. Oh yeah, and an overdose can kill you.

Nicotine also acts on the brain's reward system, but inhaling the drug helps it reach the brain even faster than it would if it were injected. Like other stimulants, nicotine can suppress appetite. (Some smokers say they love this side effect.) But smokers rapidly grow tolerant of nicotine's effects and need more and more of the drug to feel the same high.

Though high doses of caffeine can undoubtedly have unpleasant effects (ranging from irritability to the most unpleasant of all: death), small to moderate amounts can boost our mental functioning in ways researchers are now measuring.

One study showed that the equivalent of two cups of coffee can boost short-term memory and reaction time. Functional MRI scans taken during the study also found that volunteers who had been given caffeine had increased activity in the brain regions involving attention. Caffeine has also been shown to protect against age-related memory declines in older women.

COCKTAIL PARTY TIDBITS

❖ Three-quarters of the caffeine we ingest comes from coffee. Try to limit yourself to fewer than a hundred cups a day. That much coffee contains about ten grams of caffeine, enough to cause fatal complications.

❖ One of fiction's most famous stimulant users is the great capercracker Sherlock Holmes. Many of the detective's capers include descriptions of the relief he found in injecting cocaine. It must be tough to make sure justice is done.

DEPRESSANTS

Depressants are drugs that slow the central nervous system down, reducing motor activity and tension and wiping inhibitions away. Alcohol is by far the most commonly used depressant drug. According to the Centers for Disease Control, more than 60 percent of American adults reported in 2004 that they had a drink in the previous year.

Alcohol interacts with most areas of the brain, but especially the cerebellum and the cortex, causing motor and cognitive impairments. It reduces brain activity in areas involved in judgment and decision-making, creates slurred speech, and can permanently damage the brain and liver. Over the long term, alcohol use can lead to memory problems and a reduction in gray matter.

Opiates are drugs that are derived from the seed of the poppy plant and are cultivated mainly in Asia. (Poppies, and opium, are big business in Afghanistan—and have been a worsening problem since the United States ousted the Taliban leadership in 2001.) Opiates, a class of drugs that includes heroin and morphine, mimic the action of the body's natural painkillers, reducing painful sensations and increasing pleasurable ones. They work their magic in the brain's limbic system, brainstem, and spinal cord, among other places.

Barbituates are a group of depressants that were originally used as prescription sleep aids. They depress activity in the part of the brain stem that keeps us alert, causing drowsiness. In high doses they can slow or even stop breathing and result in coma or death.

ON THE FRONTIER

Regularly using opiates can make your brain more vulnerable to the effects of stress, according to a recent study. When we're stressed, certain parts of the brain release neurotransmitters that trigger the body's fight-or-flight response. Chronic opiate use makes neurons in the brain's stress system hypersensitive, causing them to fire off more of these neurotransmitters than normal.

So far, research has just identified this amplified stress response in rats—now they have to replicate the findings in humans. If they do, it could explain why opiate addicts are more likely to show symptoms of post-traumatic stress disorder than nonaddicts.

COCKTAIL PARTY TIDBITS

❖ Delirium tremens describes a severe withdrawal reaction to alcohol that includes disturbing visual hallucinations. (One of the most common is a hallucination of something, like bugs, crawling all over one's body.)

❖ Jim Morrison and Janis Joplin both died of heroin overdoses.

OTHER DRUGS

And then there are the rest. Take our good friend Mary Jane, for instance. Marijuana, the infamous "gateway drug," comes from the plant *Cannabis sativa*. The active ingredient in marijuana is tetrahydrocannabinol (THC), which interacts with the cortex, cerebellum, amygdala, and hippocampus. It produces feelings of relaxation and pleasure, and decreases concentration, memory, and dexterity. It also increases appetite, giving you the munchies. In some people, marijuana can create a sense of anxiety or panic.

Many studies have explored the potential long-term effects of marijuana use on the brain, but there is no clear scientific consensus about the results. Either way, because the drug is usually smoked, it can certainly damage the lungs. Marijuana also has pain-reducing effects, leading to interest in using the drug medicinally.

Hallucinogens—including LSD, PCP, and mescaline—alter a user's sense of reality. Many are naturally occurring, present in mushrooms and other plants. These drugs produce their effects by altering activity in the cortex, thalamus, and brain stem and influencing production of the neurotransmitter serotonin. A hallucinogenic trip may also feature intense emotion, drastic mood swings, and synesthesia. (Some hallucinogens cause trips that can last for days.)

Hallucinogens were immensely popular among the U.S. countercultural movements of the '60s and '70s. Some scientists believe that hallucinogens may be promising for treating disorders ranging from headaches to obsessive-compulsive disorder, but the use of these drugs, even only in research, remains highly controversial.

ON THE FRONTIER

Ever since the club drug Ecstasy burst onto the national scene, researchers have been exploring its short- and long-term effects on the brain. A number of studies have looked, in particular, at the effects of the substance on memory. Even low doses of the drug, a recent investigation shows, can impair verbal memory. Use of Ecstasy in the long term can disrupt the ability to form new memories, and regular users of the drug report that they have trouble remembering all sorts of things in their daily lives.

COCKTAIL PARTY TIDBITS

- ❖ Evidence of marijuana use dates all the way back to the Neolithic Age (8000 to 3500 B.C.). If they were smart, our ancestors probably discovered pot right after they created fire.
- ❖ Marijuana varieties are being bred to be more and more potent, meaning that they contain higher concentrations of THC than they used to.

NEUROTOXINS

All the drugs discussed so far are neurotoxins, meaning they mess directly with our nerve cells, causing all sorts of nasty effects. But they're just a small percentage of the known neurotoxins. Here are some of the worst offenders:

- **Lead.** Can cause learning disabilities in children. Try to put your foot down when it comes to letting your kids eat paint chips from old buildings or toys from China. You'll thank me later. (Some historians blame the fall of the Roman Empire on the Romans' reliance on lead pipes for drinking water.)

- **Mercury.** Can damage the nervous system. Mothers who are exposed to mercury with no obvious ill effects can still give birth to kids with serious neurological problems. (Lewis Carroll's Mad Hatter was inspired by hatters of the day, who used mercury when making their wares and frequently suffered subsequent mental illness.)

- **Carbon monoxide.** Targets and disrupts the nervous system. Invest in a carbon monoxide detector, and don't forget to change the batteries.

- **Venoms.** Many creatures, including scorpions, snakes, bees, and spiders, secrete neurotoxic venoms. The green mamba snake uses a venom that increases the release of a certain neurotransmitter and causes convulsions.

Venoms are one of the hottest things in medicine right now. Pharmaceutical companies are studying naturally occurring venoms in an attempt to create drugs that mimic their effects. It's not as crazy as it sounds. Controlled doses of otherwise toxic substances may be useful for controlling pain, reducing blood pressure, fighting cancer, and a whole range of conditions. Viva la venom.

COCKTAIL PARTY TIDBITS

❖ Botox, injected in the face to reduce the appearance of wrinkles, is actually comprised of botulinum toxin A, one of the most poisonous substances on Earth. It blocks nerve messages traveling to muscles, causing paralysis, and in anything but infinitesimal doses can be deadly.

❖ The pufferfish contains a highly dangerous neurotoxin known as tetrodoxin, which can be lethal. But that hasn't stopped the fish from becoming a delicacy, known as fugu, in Japan. The minute amount left in the flesh can make your lips tingle pleasantly, but if prepared incorrectly, diners can be poisoned and die.

CHILD ABUSE AND NEGLECT

Given how important the first decade or so of life is for proper neurological and psychological fine tuning, it's not surprising that abused children develop problems that can last a lifetime.

Victims of abuse tend to do poorly in school and be at greater risk for juvenile delinquency and substance abuse. Over the course of their lifetimes, they are also more likely to suffer from medical and psychological problems.

Abused children have chronically high levels of cortisol, a stress hormone, which seems to delay their development, making them slower walkers and learners than other kids. Studies have also shown that they have disturbances in their limbic systems and that many of their brain structures are smaller than those of children who are not abused.

It's not just active abuse that can derail development— parental neglect can be similarly devastating. Rodents who have physically affectionate mothers (rodent affection in- cludes a lot of licking and grooming) grow into adults with lower levels of anxiety and healthier responses to stress than pups who are not lavished with such attention.

Other animal research shows that separating a young creature from its mother for a long period of time not only disrupts the growth of certain neurons but also causes some brain cells to die. Children who don't form secure attachments with parents or caregivers may later have trouble forming healthy bonds with anyone else.

So put this book down and hug your kids.

A recent study suggests a neurological mechanism to explain why many abused children become abusers themselves. Rhesus monkeys who are abused by their mothers during the first month of their lives have brains that produce less serotonin. And abused infants who grow into abusive parents have lower serotonin levels than abuse victims who managed to break the cycle.

Low levels of the neurotransmitter serotonin have been linked not only with anxiety and depression but also with aggression, potentially explaining the link.

COCKTAIL PARTY TIDBITS

❖ Children who are disabled—physically or intellectually—are more likely to be victims of abuse.

❖ As many as 900,000 kids a year are victims of abuse or neglect, according to a 2000 report.

❖ More than 10 percent of women report being sexually abused by an adult when they were children.

COMBAT

THE BASICS

The United States' war in Iraq is rapidly expanding our knowledge of the effects of combat on the brain. Unsurprisingly, these effects are not positive.

One of the most devastating types of injuries that soldiers are returning home with is known as traumatic brain injury (TBI). Two-thirds of the Iraqi vets at Walter Reed Army Medical Center have these wounds. TBIs are caused by severe blows to the head or penetrating head wounds; in Iraq, they are often caused by improvised explosive devices.

The injuries can cause loss of consciousness and amnesia. In the months that follow, other symptoms may develop, including headaches, insomnia, and lasting cognitive difficulties. Paradoxically, the increase in TBIs in returning soldiers can be attributed to improvements in body armor—in past conflicts, many of the soldiers who sustained similar injuries wouldn't have survived.

The effects of traumatic brain injuries are often compounded by post-traumatic stress disorder. Brain injuries and PTSD can make readjustment to life back home difficult. For instance, soldiers who suffered mild traumatic brain injuries and PTSD were more likely to miss work than veterans with other types of combat injuries.

TBIs can have long-term effects. Researchers looked at vets of the Vietnam War who had suffered penetrating head injuries. As these men got older, their mental function slipped—and it did so much faster than the cognitive skills of men who hadn't sustained such trauma.

The glimmer of good news is that the men who had the highest levels of intelligence before their head injuries had the slowest declines. This cognitive deterioration is different from dementia, the scientists say, and will likely also be found in Iraq vets who are suffering from similar injuries.

COCKTAIL PARTY TIDBITS

❖ Doctors often discussed World War I–related "shell shock," a euphemism for what we now know as PTSD.

❖ Between 15 and 20 percent of U.S. soldiers who survived the Vietnam War suffered from a head injury.

❖ Military law stipulates that soldiers who attempt suicide can be prosecuted and even imprisoned. Yeah, that's a good idea. Prison is just the thing to help veterans recover.

VIDEO GAMES

Video games could save your life. Surgeons who spend at least a few hours a week playing video games make one-third fewer errors in the operating room than nongaming doctors. Indeed, research has shown that video games can improve mental dexterity, boosting hand–eye coordination, depth perception, and pattern recognition. Gamers also have better attention spans and information-processing skills than the average Joe. When nongamers agree to spend a week playing video games (in the name of science, of course), their visual perception skills improve. And strike your notions of gamers as outcasts: One researcher found that white-collar professionals who play video games are more confident and social.

Of course, we can't talk about the effects of video games without mentioning the popular theory that they are responsible for increasing real-world violence. A number of studies have reinforced this link. Young men who play a lot of violent video games have brains that are less responsive to graphic images, suggesting that these gamers have become desensitized to such depictions. Another study revealed that gamers had patterns of brain activity consistent with aggression while playing first-person shooter games.

This doesn't necessarily mean that these players will actually be violent in real life. The connections are worth exploring, but so far the data do not support the idea that the rise of video games is responsible for increased youth violence.

ON THE FRONTIER

Video games activate the brain's reward circuits, but do so much more in men than in women, according to a new study. Researchers hooked men and women up to functional MRI machines while the participants played a video game designed for the study. Both groups performed well, but the men showed more activity in the limbic system, which is associated with reward processing. What's more, the men showed greater connectivity between the structures that make up the reward circuit, and the better this connection was in a particular player, the better he performed. There was no such correlation in women. Men are more than twice as likely as women to say they feel addicted to video games.

COCKTAIL PARTY TIDBITS

❖ Video games are a ten-billion-dollar industry in the United States.

❖ In 2003, a sixteen-year-old boy shot and killed two police officers and a police dispatcher. Two years later, the families of the victims filed suit against the company that made the massively popular video game Grand Theft Auto. The lawsuit alleges that the perpetrator was inspired by his obsession with the controversial game.

MUSIC

THE BASICS

When you turn on Queen's *Greatest Hits,* the auditory cortex analyzes the many components of the music: volume, pitch, timbre, melody, and rhythm. But there's more to music's interaction with the brain than just the raw sound. Music can also activate your brain's reward centers and depress activity in the amygdala, reducing fear and other negative emotions.

A highly publicized study suggested that listening to Mozart could boost cognitive performance, inspiring parents everywhere to go out and buy classical CDs for their children. The idea of a "Mozart effect" remains popular, but the original study has been somewhat discredited, and any intellectual boost that comes from listening to music seems to be tiny and temporary. But music does seem to possess some good vibrations. It can treat anxiety and insomnia, lower blood pressure, soothe patients with dementia, and help premature babies gain weight and leave the hospital sooner.

Music training can bolster the brain. The motor cortex, cerebellum, and corpus callosum (which connects the brain's two sides) are all bigger in musicians than in nonmusicians. And string players have more of their sensory cortices devoted to their fingers than those who don't play the instruments.

There's no agreement yet on whether musical training

makes you smarter, but some studies have indeed shown that music lessons can improve the spatial abilities of young kids.

ON THE FRONTIER

Music lessons and practice during childhood increase the sensitivity of the brainstem to the sounds of human speech. According to a recent study, the brainstem is involved in very basic encoding of sound, and lots of exposure to music can help fine-tune this system, even in kids without particular musical gifts.

So buck up, tone-deaf children of the world! Think of it like eating vegetables: Chewing on that clarinet is good for you.

COCKTAIL PARTY TIDBITS

* The auditory cortex is activated by singing a song in your head. The visual cortex is activated by merely imagining a musical score.
* Playing classical and soothing music can increase the milk yield of dairy cows.

MEDITATION

Forget apples. If an entire ream of scientific studies is to be believed (and such studies usually are), an *om* a day could keep the doctor away. Meditation, or the turning of one's mind inward for contemplation and relaxation, seems to help all sorts of conditions—anxiety disorders, sure, but it can also reduce pain and treat high blood pressure, asthma, insomnia, diabetes, depression, and even skin conditions. And regular meditators say they feel more at ease and more creative than nonmeditators.

Researchers are now illuminating the actual brain changes caused by meditation, by sticking meditators into brain imaging machines. Regular meditators show a variety of brain changes while meditating. For one, although the brain's cells normally fire at all different times, during meditation they fire in synchrony.

Expert meditators also show spikes of brain activity in the left prefrontal cortex, an area of the brain that has generally been associated with positive emotions. And those who had the most activity in this area during meditation also had big boosts in immune system functioning.

Meditation can increase the thickness of the cerebral cortex, particularly in regions associated with attention and sensation. (The growth doesn't seem to be because the cortex grows new neurons, though—it appears to be because the neurons already there make more connections, the number of support cells increases, and blood vessels in that area get bigger.)

ON THE FRONTIER

Meditation can increase focus and attention, improving performance on cognitive tasks. Researchers spent three months training volunteers in the practice of Vipassana meditation, which focuses on minimizing distractions.

Then, the volunteers were asked to perform a task in which they had to pick a few numbers out of a stream of letters. People who had undergone the training in meditation were much better at identifying numbers that briefly flashed onto a computer screen. They also seemed to be able to do this without exerting as much mental energy.

COCKTAIL PARTY TIDBITS

❖ Monks who take part in these scientific studies have typically spent more than ten thousand hours in meditation. That's more than a year.

❖ The Dalai Lama was a distinguished speaker at the Society for Neuroscience's annual conference, the world's largest gathering of brain researchers.

A MIND OF ONE'S OWN: INDIVIDUAL BRAINS

GENES AND THE BRAIN

The '90s might as well have been the "gene for" decade. Every time you read the news, it seemed, another gene had been discovered: the "gene for autism," the "gene for anxiety," the "gene for aggression," and so on. But things are a lot more complicated than that, especially when it comes to the genetics of the brain and behavior.

We've all heard of "nature versus nurture," but the idea that it's one or the other is far too simple. It's the interaction of the genes we inherit with the life experiences we accumulate that make us who we are. When newspaper stories start talking about a "gene for aggression," for instance, what they mean is not that anyone with the gene becomes aggressive or that anyone who is aggressive has the gene. It doesn't even mean that someone who is aggressive and has the gene is aggressive *because* they have the gene.

It's sort of like baking a cake—you need lots of ingredients and there are lots of different recipes. Mix together a genetic variant, an abusive childhood, lots of school bullying, and lax gun control laws and voilà: You have a school shooting. This is generally how most of the genes we've discussed work. A gene is simply one ingredient.

There are some exceptions—Huntington's disease, for instance, depends solely on inheriting a particular genetic mutation—but very, very few. That should be liberating to hear: Genes, in almost every case, simply are not destiny.

ON THE FRONTIER

Genomic technology has improved greatly in recent years, and scientists are using it to identify and sequence genes that influence the size and development of the brain. One of the flat-out coolest applications to emerge from the data is the Allen Brain Atlas, an interactive, three-dimensional map of all the genes that are active in the brain. The atlas allows researchers to investigate areas of interest and see what particular genes are "turned on" in a particular part of the brain.

The tool could help identify stretches of DNA that might be involved in disorders ranging from autism to Alzheimer's.

COCKTAIL PARTY TIDBITS

- You probably already know what eugenics is: a theory of improving society and the human race by allowing only those with the best DNA to reproduce. But have you heard of euthenics? This countertheory suggests that humanity would be better served by improving housing, education, sanitation, and more. Plus it doesn't recommend sterilizing people against their will.

- We *Homo sapiens* have somewhere in the neighborhood of twenty to twenty-five thousand genes.

INTELLIGENCE

Intelligence is like pornography—nearly impossible to define, but we all know it when we see it. In general, though, intelligence refers to our ability to use reasoning to solve problems. The most well-known assessment of intelligence is the Stanford–Binet test, which assesses quantitative reasoning, spatial skills, verbal ability, and more. The test then generates an intelligence quotient, or IQ. A score of 100 is considered average; 140 is genius-level.

Despite their longstanding use, IQ tests are controversial and not terribly politically correct—many people argue that they measure narrow kinds of book smarts, that they're biased against minorities, and that IQ scores are not necessarily related to success. (Indeed, there is also a long-running and controversial debate over why minorities tend to perform worse on the Stanford–Binet test, but we don't have the space—or the lawyers—to go into it all here.)

Researchers have attempted to measure all sorts of alternative kinds of intelligence. The most notable is emotional intelligence, or EQ, which refers to someone's ability to understand and manage social relationships.

Intelligence is largely inherited. Sure, environmental factors make a difference—plenty of stimulation might help your kid eke out a few extra IQ points, neglecting him might do the opposite—but research has repeatedly suggested that our genes largely determine our rough position on the intelligence scale.

Intelligence may be genetic, but don't tell your kids that. Junior high school students who think intelligence can be developed do better in school than those who think it's fixed.

All the students started out with similar levels of accomplishment in math. But over the course of two years the students who said they believe intelligence was a trait that could be nurtured outperformed the others. The scientists say they believe that kids who think intelligence can be developed also believed more in the importance of learning and effort.

COCKTAIL PARTY TIDBITS

* Average intelligence scores have been increasing rapidly over the last century or so. This phenomenon is known as the Flynn effect.
* The highest IQ on record, 228, belongs to Marilyn vos Savant, who then did an incredibly stupid thing when she agreed to write a regular column for mind-numbing *Parade* magazine.

PERSONALITY

There's not enough space to list or discuss the various theories about where personality comes from and how we can best describe it, but there are some useful things we can say about personality in general.

There are many different ways of breaking up and classifying personalities, but one widely accepted system identifies the "big five" dimensions of personality: extroversion (would you rather spend your evening at a big party or a quiet wine bar?), openness (would you be willing to date a guy who has had his chest waxed?), agreeableness (do disagreements make you squirm?), conscientiousness (do you regularly miss deadlines?), and neuroticism (does criticism make you want to crawl under the covers and hide?).

Though they hardly seem like real people yet, infants already exhibit a wide array of personalities and temperaments. Some traits seem more heritable than others, including shyness, aggressiveness, and excitability. Scientists are still trying to work out the neural basis of personality, but it's a complex problem.

What we do know, however, is that injuries to the brain can drastically change someone's personality. Phineas Gage was a friendly and well-mannered man who, after damage to his frontal lobes, became vulgar, impulsive, and self-centered, to a degree that his friends said they no longer even knew him. Similar cases dot the medical literature.

ON THE FRONTIER

It's not uncommon to hear that certain personality traits are related to certain diseases, that driven people with "type A" personalities, for instance, have more heart disease. But a host of studies suggests that the relationship between personality and health is anything but simple.

It was often suggested that certain personality traits—particularly neuroticism and extroversion—could even make people more susceptible to cancer. But a study of thirty thousand Swedish twins (wouldn't you love to see them all lined up?) revealed that there is no link whatsoever between cancer and personality.

COCKTAIL PARTY TIDBITS

❖ One of the best known personality assessments was developed by Hermann Rorschach, who had the idea that essential characteristics about someone's temperament could be determined by analyzing the images they saw in a series of ambiguous splotches. The so-called Rorschach inkblot test still looms large in the popular imagination.

GENDER AND THE BRAIN

Searching for differences between the brains of men and women remains controversial. But neuroscientists are finding undeniable distinctions. Men's brains are about one hundred grams heavier than women's, even when you take men's larger body size into account. But misogynists take note: Women's brains have more densely packed neurons and more gray matter.

Cognitive tests have also shown that men and women tend to excel at different tasks. Women, as a group, outperform men on tests of verbal skills, while men do better at tasks requiring spatial reasoning. There are also well-documented differences in how the two sexes navigate. Females tend to use landmarks to find their way, while men rely on estimations of distance and cardinal directions.

Of course, all these findings come with plenty of caveats. Chief among them: These findings are merely averages across large populations, and there is plenty of overlap among men and women. That means you can't assume a woman is incapable of spatial reasoning just because she lacks a Y chromosome.

These brain differences are also likely to be a combination of genetics, hormones, and environment, and researchers are still busy untangling the influence of each of these factors.

Researchers are trying to explain why women seem more susceptible to drug addiction than men are.

Recent studies of rodents show that female rats were also more likely to self-administer cocaine when their levels of circulating estrogen were high. Estrogen appears to activate brain regions involved in reward processing, the researchers say. Other work has suggested that women may have genetic predispositions that increase the activity of the brain's pleasure centers and thus explain their higher vulnerability to cocaine addiction.

COCKTAIL PARTY TIDBITS

❖ Conventional wisdom has long held that the gift of gab is for those with two X chromosomes. But a new analysis of decades of studies on gender and talkativeness reveals a small—but significant—trend that men are more talkative. This difference is especially pronounced when men are talking to their spouses or to strangers, whereas women are likely to be slightly more talkative than men when talking to their children or college classmates.

SEXUAL ORIENTATION

Much recent attention has been focused on the basis of human sexual orientation. Though many findings are contradictory, it seems clear that homosexuality is a result of both genetic and environmental components. ("Choosing" to be gay is not one of those factors.)

One of the most widely documented contributions to homosexuality is male birth order; the more older brothers a man has, the more likely he is to be gay. Each older brother increases by 33 percent a man's chance of being gay.

It has been hypothesized that this so-called fraternal birth-order effect is a result of an immune reaction in the mother. A male fetus, it is thought, prompts a mother to mount an immune response, in which her antibodies attack the substances that helps a fetus's brain become masculine instead of feminine. This maternal immune response may be cumulative—the more male babies she's had, the more strongly her system will attack these masculizing substances in a subsequent male fetus.

Indeed, some research has found that certain features of gay men's brains seem more similar to the characteristics of female brains than male ones. Other studies suggest that there is at least a modest genetic contribution to homosexuality in men; several indicate that one of these genes may be located on the X chromosome, which men inherit only from their mothers. As for lesbianism, there is much less research on its origins and biological basis. Another glass ceiling for women to break through.

ON THE FRONTIER

An established body of research has shown that men and women differ in their ability to learn and remember spatial information (see the previous topic for more). Now, a new study shows that gay men more closely resemble women than men when navigating. Gay men and straight women both took longer than straight men to complete certain spatial tasks, and gay men, like women, tended to use landmarks to navigate. Next up: Will gay men ask for directions?

COCKTAIL PARTY TIDBITS

❖ Homosexuality used to be classified as a disorder by what's known as the Diagnostic and Statistical Manual of Mental Disorders (DSM), the mental health bible that lists the diagnostic criteria for all recognized psychiatric problems. It was removed from the DSM in 1973, after which outlawing same-sex marriage became the technique of choice for marginalizing gays and lesbians.

❖ There are more than 450 species of animals (that we know of) that engage in some form of homosexual behavior.

SEX HORMONES

The sex hormones estrogen and testosterone don't just have reproductive functions—both substances influence the workings of the brain. Even though estrogen is a female sex hormone and testosterone is thought of as a male one, both substances are present in men and women.

Estrogen boosts neuron growth and connectivity, which means it probably aids in information processing. It also influences mood and improves learning and memory. It's thought that the estrogen deficits in menopausal and postmenopausal women might contribute to memory problems, dementia, and more.

However, given the surprisingly negative results of a long-term study of estrogen replacement therapy (it found that estrogen replacements actually increased the incidence of all sorts of medical problems, including Alzheimer's), the future potential of estrogen supplements to stave off memory problems remains uncertain.

Testosterone has some similar effects in the brain, boosting learning and memory. Researchers have found that testosterone supplements improve performance on certain tests of memory and that men with Alzheimer's have lower levels of the sex hormone than their healthy peers. Much research has also focused on the role of testosterone in aggression, and both high and low levels of the hormone have been linked to aggressive behavior.

ON THE FRONTIER

Depression and anxiety disproportionately affect women, and research suggests that estrogen may play a role in this. Elevated estrogen levels can make the brain more susceptible to the effects of stress. Scientists have discovered that moderate levels of stress can interfere with female rats' performances on tests of short-term memory. Male rats did not show this sensitivity.

What's more, the stress affected the female rats only if they happened to have high estrogen levels at the time. (Estrogen levels change throughout the reproductive cycle.) The findings have implications not only for why women are more vulnerable to certain mental illnesses but also why these sex differences first become apparent during puberty.

COCKTAIL PARTY TIDBITS

❖ Women with lower levels of estrogen are more sensitive to pain.
❖ The brain can convert testosterone into estrogen, potentially confounding studies that try to separate the effects of these hormones. But researchers are working to untangle them.
❖ Men who have naturally high levels of testosterone tend to have more children, but don't live as long.

FETAL BRAIN

Brain development begins in earnest a few weeks after conception. Around this time, a flat neural plate starts to form
and soon grows into what's called the neural tube. Over the
course of fetal development, the neural tube eventually grows
into the various subdivisions and structures of the brain.

Neurons are generated at an astonishing pace in the fetal
brain. As many as 250,000 neuroblasts (pre-neural cells) are
born per minute in the neural tube. These huddled masses
of neurons then migrate to other locations in the developing
brain, each neuron taking up residence in what will be its fi-
nal location. Glial cells guide and assist these traveling nerve
cells.

Once the neurons arrive at their destinations, they begin
forming connections nearby neurons and begin to differentiate themselves—for instance, those that make their home in
the auditory cortex become auditory neurons.

During later stages of development, the fetal brain starts
to get rid of some dead weight. Neurons that aren't getting
enough signals die off and connections that aren't getting
enough use wither. This cleansing is a normal—and vital—
part of development, helping to refine the brain.

Oh, and one more thing: The fetal brain is a delicate
thing. That means mothers can start screwing up their kids

even before they're born. Smoking, drinking alcohol, and even just plain not eating right can all interfere with fetal brain development, causing permanent damage.

Bad news for all you moms-to-be who want to mellow out a little: Smoking pot can harm your baby's brain.

Research reveals that fetal brain development is governed, in part, by molecules in the brain that resemble THC, the active substance in marijuana. These molecules help the neurons in the developing brain establish connections. That means that any THC that gets into the fetal brain (from a pot-smoking mom) could disrupt this process and lead to improper mental wiring.

COCKTAIL PARTY TIDBITS

- ❖ Before the massive neuron die-off begins, the fetal brain has *twice* as many nerve cells as the adult brain.
- ❖ Expectant mothers don't just need to worry about undernourishment. Too *much* of certain vitamins, notably vitamins A and D, can also interfere with proper fetal brain development. Mothers these days just can't catch a break.

INFANT BRAIN

Kids are not little adults, and their brains are clear evidence of that. Babies are born with some brain functions already working (reflexes, for instance), but their brains still have many years of serious development ahead of them. Children have many neurons when they're born, and these neurons spend their early years growing at a furious pace, stretching out axons and dendrites like trees sprouting new branches.

But not all of these branches will survive into adulthood. Just like a tree, a healthy brain requires pruning away weak and dead branches. The connections that get used during childhood—during a rollicking game of peek-a-boo or a meal of strained peas, for example—get stronger, while those that don't gradually wither and die.

This process of brain pruning fine-tunes the brain's connections. But it also means that kids require lots of sensory input during development. If children miss out on stimulation, the damage may be irreversible.

Evidence of this comes from a misguided treatment that was once used on children who were born cross-eyed. These infants would routinely have one of their eyes covered with a patch while doctors waited for the eye muscles to develop enough to be operated upon. But researchers found that after the patch was removed, the children never learned how to see out of the eye that had been covered—without visual input, the proper nerve pathways never developed.

ON THE FRONTIER

Even infants have a sense of quantity, according to a recent study. Scientists outfitted the infants with brain electrodes and presented images of objects.

Occasionally, the number of objects in the image would change. When this happened, the infants would show a flurry of brain activity. This activity, it turns out, occurred in the same brain regions that process numerical information in adults. The research suggests that the neural pathways for number are laid down early in the course of brain development. If only long division was so natural.

COCKTAIL PARTY TIDBITS

❖ An infant's brain grows faster than his body, *tripling* in size by the time his first birthday rolls around.

❖ Your little bundle of joy may not be able to help you with the crossword puzzle, but at the age of three, her brain is twice as active—that is, it uses twice the amount of fuel—as yours.

❖ The brain circuits involved in forming and storing memories don't mature until about three or four years of age, explaining why we have few memories from earlier in our lives.

TEENAGE BRAIN

We've all done stupid things, especially when we were teenagers: Stolen a tube or two of lip gloss, tried substances we shouldn't have, used the letter *z* to make the word *boy* plural.

So perhaps it's not surprising that research is increasingly revealing that teenagers literally have minds of their own. Neither children nor adults, adolescents have brains that make them uniquely susceptible to risk-taking and the consequences of such behavior. Though some brain areas mature during childhood, the prefrontal cortex—which is involved in foresight and self-control, among other things—doesn't finish developing until the early twenties.

Teenagers trying to behave are also fighting against high levels of the neurotransmitter dopamine, which peaks in the adolescent brain. Dopamine is intimately connected with reward-seeking behavior and also with reinforcing pleasurable activities. Too little self-control and too much dopamine? Sounds like a cocktail for risky behavior.

What's more, the adolescent brain is uniquely sensitive to the consequences of these "risky behaviors." (Do I sound like your high school gym teacher yet?) Studies show that teen brains form stronger connections to rewarding stimuli (drugs, for instance) and that these associations last longer than they do in adult brains.

This means that drug and alcohol addiction may be harder to treat in teenagers and that relapse may be more likely. If only the "my-still-developing-frontal-lobe-made-me-do-it" defense would hold up in juvenile court.

ON THE FRONTIER

Recent studies have shown that, during adolescence, the brain starts to lose some of its gray matter. This loss, however, seems to be a good thing, accompanied by improved mental and cognitive functioning. Scientists suspect that the loss of gray matter actually means the brain is becoming more efficient by eliminating unnecessary connections. It may also mean that more neurons are becoming coated in myelin which increases signaling speed.

COCKTAIL PARTY TIDBITS

- ❖ Adolescents are at particular risk for a number of mental illnesses, particularly mood disorders. At this very moment, between 10 and 15 percent of teens are struggling with the symptoms of depression.
- ❖ Love hurts: Adolescents who are involved in romantic relationships are more likely to develop depression or problems with alcohol, and girls in relationships are at more risk than boys.

PARENTAL BRAIN

We all know that pregnant women go through changes. But it's not just their bellies or their moods—becoming a parent causes a variety of neurological and behavioral changes. Good parenting behaviors are associated with two hormones in particular: cortisol and prolactin.

Mothers who are most responsive to the cries of babies have the highest levels of cortisol (yes, that's the "stress hormone"). And reducing a mother's level of prolactin, the hormone that causes the breasts to produce milk, can turn off her natural maternal instincts.

Things really get interesting when it comes to daddy dearest. Research is increasingly showing that fatherhood comes with biological and neurological changes, just as motherhood does.

Male marmosets also pack on pounds during their mates' pregnancy, gaining as much as 20 percent of their body weight. This weight gain is thought to prepare the dudes for the extra work (and energy) tending to baby requires.

Human dads-to-be also have high levels of cortisol and prolactin, just as their female partners do. In fact, dads with elevated levels of prolactin express more concern when they hear the sound of a crying infant.

Having little ones seems to enhance the prefrontal cortices of marmoset males, who have more neuron connections in this region after they become dads.

The prefrontal cortex is important for planning, organization, and other executive functions that every parent needs. The neurons in this region also have more receptors for vasopressin—another parenting hormone—in marmoset dads than in bachelors. It's all neurological insurance, perhaps, for making sure that father really does know best.

COCKTAIL PARTY TIDBITS

❖ A man's level of testosterone drops by approximately one-third in the days and weeks after he becomes a parent. It's thought that this drop makes dads more likely to stay at home and nest and less likely to go out and find new sexual partners.

❖ How strong is the maternal instinct? Momma rats that are allowed to choose between tending to their young pups and taking a hit of cocaine choose their offspring. Now that's love.

AGING BRAIN

With age comes wisdom? Screw that. With age come neuro-degenerative diseases, and scientists are learning more about them every day. But research is also beginning to reveal the changes that even the healthiest of brains undergo as they age. As we get older, our brains begin to lose gray matter as neurons atrophy. (But unlike the gray matter loss that comes during the teenage years—and seems to reflect a brain that is becoming more efficiently organized—the loss of gray matter that accompanies aging is not a good thing.)

What's more, the myelin coating on many of our neurons begins to deteriorate. That means communication among neurons may become less reliable. The number of synapses present in the cerebral cortex also declines with age; it is the evaporation of these crucial links that may underlie the memory loss associated with old age. But it is still unclear exactly how closely linked this physical deterioration is with cognitive decline—intelligence may remain stable in the face of these and other brain changes.

But take heart, those approaching (or even well past) the hill! Even the oldest brains are still capable of incredible feats. Research shows that aging rats that are exposed to stimulating environments are able to form new synapses, just as young'uns do. Intellectual physical exercise can both keep aging brains feeling spry. So embrace that daily Sudoku puzzle!

ON THE FRONTIER

Genes may make certain brains more susceptible to the wear and tear of aging. For instance, researchers have discovered a gene that seems related to age-related changes in the hippocampus, a brain structure important in memory.

People with normal copies of this gene tend to be better equipped to stave off age-related declines in the hippocampus. But those with a mutated version of the gene are more likely to have a dramatic deterioration in the hippocampus, and its activity, as they age.

COCKTAIL PARTY TIDBITS

❖ Don't get too attached to those neurons—your brain will lose 10 percent of its weight over your lifetime.

❖ In a study of more than six hundred people, scientists found that those with positive perceptions of aging lived an average of seven and a half years longer than those who had less rosy views of old age. Further analysis revealed that this bump in longevity was not merely because optimistic patients were more likely to have money, friends, or good health. The researchers conclude that rosy-colored glasses alone are enough to prolong life.

EVOLUTION

Our brains have come a long way, thankfully. (You wouldn't want one of evolution's first drafts.) The human brain isn't the biggest brain out there (that distinction belongs to the sperm whale), but it is the largest proportional to body size. The modern human brain weighs 1,350 grams, while the brain of the chimp, our closest relative, weighs in at only 400. (And chimps are only a bit smaller than we are.)

Indeed, research shows that after the human and ape evolutionary trees parted ways about five and a half million years ago, our brains underwent a rapid expansion. But this growth wasn't indiscriminate. The cerebral cortex grew more than other areas, and over the course of primate evolution, more volume was added to the front of our brains—the areas that we associate with personality, decision-making, morality, and the like.

But what spurred these changes? Evolutionary questions like these are, by definition, historical, so it can be difficult or impossible to determine an unequivocal answer. Some classic theories suggest that tool use and language spurred our evolutionary brain boom. Others suggest that diet had something to do with it—that better and more nutritious food made it possible for our brains to bloom. Or that humans' big groups and community structure required brains big enough for parsing a complex social world. The answer may be some combination of these factors or something else we haven't even thought of yet.

ON THE FRONTIER

Don't feel quite smart enough yet? Just wait a while. The human brain is still evolving. Scientists have identified two different genes that help regulate the size of our brains. And these genes have been evolving rapidly in modern humans. New variations of these genes arose relatively recently, but have become more and more common in humans. Given how fast these genetic variants have spread, it's likely that they produce some sort of advantage for people lucky enough to possess them. But not all scientists agree that the variant improves brain function—it's all still a theory.

COCKTAIL PARTY TIDBITS

- ❖ Okay, I know I said that humans have the biggest brains in proportion to body size, but some scientists debate that. It seems that shrews might have a slightly larger brain-to-body ratio.
- ❖ Einstein's brain weighed in at only 1,230 grams—considerably less than average for humans. On the other hand, a honeybee's brain is a mere milligram, so Einstein clearly had bees beat.

ANIMAL MINDS

As fast as we can come up with traits that distinguish us from the rest of the kingdom of the beasts, scientists find animal species that share those characteristics. Some of the higher functions animals seem capable of include:

❖ **Tool-making.** The ability to create and use tools was once considered unique to humans, or at least to higher primates. But tool-making is now well documented in many species, including dolphins, which use sponges to protect their snouts while foraging, and even crows, which have proven themselves remarkably capable of designing novel tools to help them grab pieces of meat.

❖ **Culture.** Animals not only make tools, but they pass this knowledge down, and different communities of animals of the same species have unique habits and behaviors.

❖ **Self-awareness.** Self-awareness has long been considered a hallmark of humanity. But recent research suggests that even elephants are capable of passing the mirror test for self-awareness. When researchers put a mark of paint above an elephant's eye and then presented the pachyderm with a mirror, the elephant reached up with her trunk to touch the mark above her eye, proving that she recognized herself. Dolphins and chimpanzees have also passed this test.

Some animals are able to process quantitative information. A study of lemurs, primates that were long considered to be relatively low on the intelligence totem pole, showed that they can distinguish between different quantities of fruit.

Other research has shown that dolphins can identify which image in a set has the fewest number of dots, indicating that the mammals understand the somewhat sophisticated principle of one set being "numerically less" than another. And chimps—our dear, overachieving cousins—even have better short-term memories for numbers than we do. *Harrumph.*

COCKTAIL PARTY TIDBITS

- ❖ Dolphins use unique sounds to identify individuals, much in the way we humans use names. They name themselves with whistles, calling out their own names in the wild, and research has shown that other members of their social group remember and recognize these unique "names."
- ❖ Elephants are big brained (not to mention big boned) in general, but their hippocampi are especially large and wrinkled. This might explain the beasts' famous memory skills.

CONSCIOUSNESS

According to *Merriam-Webster,* consciousness . . . no, just kidding. Regardless of whatever the dictionary might have to say, you'd have a hard time finding anyone who really knows what consciousness is (and if you do, you might as well send them directly to the Nobel prize committee).

The brain is a purely biological organ; its workings are nothing more than a collection of natural, physiological processes. But somehow, all that neuron signaling, all those electrical changes, all those busy chemical messengers—somehow give rise to a sense of self.

Our neurons prompt our behaviors, but each and every one of us believes that there is a subjective "I" that is driving the car. How do we get from neurons to "I"? Scientists now know a lot about different states of consciousness (whether we're awake, delirious, in comas, and so on), but the big question—how the workings of our neurons produce the experience of individual identity—remains a mystery. We're not even really sure how to formulate this question scientifically, and some scientists say we may never be able to do so. (And it's not often you'll get a scientist to acknowledge that a research question might be eternally beyond our grasp.)

A virtual reality program might shed light on the conscious experience by disrupting individuals' senses of their own bodies. The program allows participants to watch 3D representations of themselves being stroked by a brush as they themselves are stroked in real life.

Those stroked in time with the 3D representations are more likely to sense that their bodies are located in the virtual space. The results show that researchers were able to disrupt individuals' sense of self by manipulating their sense of where their bodies were located in space.

COCKTAIL PARTY TIDBITS

❖ If we haven't solved the problems of consciousness yet, it's not for lack of attention. Consciousness has been deeply debated throughout history by such luminaries as Cicero, René Descartes ("I think, therefore I am"), and Friedrich Nietzsche.

❖ A remarkable phenomenon known as blindsight reveals how much of our brains' work never reaches the level of consciousness. Blindsight can occur in people who have damage to their visual processing centers. Researchers have held up images in these patients' blind spots; though the individuals say they cannot see anything, when researchers force them to guess whether the image is red or blue, for instance, they often do so correctly.

APPENDIX A

COMMONLY PRESCRIBED PSYCHIATRIC MEDICATIONS

Brand Name	Generic Name	Commonly Prescribed for
Abilify	aripiprazole	schizophrenia
Adderall	amphetamine salts	ADHD
Ambien	zolpidem tartrate	insomnia
Antabuse	disulfiram	alcoholism
Ativan	lorazepam	anxiety disorders
Celexa	citalopram	depression
Clozaril	clozapine	schizophrenia
Compazine	prochlorperazine	schizophrenia
Concerta	methylphenidate	ADHD
Cymbalta	duloxetine	depression
Depakote	divalproex sodium	bipolar disorder
Effexor	venlafaxine	depression, OCD, panic disorders
Halcion	triazolam	insomnia
Haldol	haloperidol	schizophrenia
Klonopin	clonazepam	anxiety disorders, bipolar disorder
Lamictal	lamotrigine	bipolar disorder, epilepsy
Lexapro	escitalopram	depression

Brand Name	Generic Name	Commonly Prescribed for
Lithonate	lithium	bipolar disorder
Lunesta	eszopiclone	insomnia
Methadone	dolophine	opiate addiction
Nardil	phenelzine sulfate	depression, panic disorders
Neurontin	gabapentin	seizures
Paxil	paroxetine	depression, OCD, panic disorders
Provigil	modafinil	narcolepsy
Prozac	fluoxetine	depression, OCD, panic disorders
Risperdal	risperidone	schizophrenia
Ritalin	methylphenidate	ADD
Seroquel	quetiapine fumerate	bipolar disorders, schizophrenia
Sonata	zaleplon	insomnia
Thorazine	chlorpromazine	schizophrenia
Topamax	topiramate	epilepsy
Valium	diazepam	anxiety disorders
Wellbutrin	bupropion	depression
Xanax	alprazolam	anxiety disorders
Zoloft	sertraline	depression, OCD, panic disorders
Zyprexa	olanzapine	bipolar disorder, schizophrenia

APPENDIX B

MENTAL HEALTH RESOURCES

National Institute of Mental Health
www.nimh.nih.gov

Mental Health America
www.nmha.org

MedlinePlus
www.nlm.nih.gov/medlineplus/mentalhealth.html

*Substance Abuse and Mental Health Services
    Administration's National Mental Health Information
    Center*
mentalhealth.samhsa.gov/default.asp

The Dana Foundation
www.dana.org

American Psychological Association
www.apa.org

American Psychiatric Association
www.psych.org

American Academy of Child & Adolescent Psychiatry
www.aacap.org

Federation of Families for Children's Mental Health
www.ffcmh.org

National Alliance Mental Health
www.nami.org

National Mental Health Consumers' Self-Help Clearinghouse
www.mhselfhelp.org

NARSAD (previously known as *National Alliance for
    Research on Schizophrenia and Depression*)
www.narsad.org

RECOMMENDED READING

The brain is so important, so complex, and so fascinating that this short volume can't even begin to do it justice. If what you've read here inspires you to learn more, check out one of these acclaimed books.

The Secret Life of the Brain by Richard Restak
   How the Mind Works by Steven Pinker

Synaptic Self: How Our Brains Become Who We Are by
   Joseph LeDoux

*Phantoms in the Brain: Probing the Mysteries of the Human
   Mind* by V. S. Ramachandran and Sandra Blakeslee

*Why Zebras Don't Get Ulcers: The Acclaimed Guide to Stress,
   Stress-Related Diseases, and Coping* by Robert M. Sapolsky

*In Search of Memory: The Emergence of a New Science of
   Mind* by Eric Kandel

Stumbling on Happiness by Daniel Gilbert

The Language Instinct: How the Mind Creates Language by Steven Pinker

The Lucifer Effect: Understanding How Good People Turn Evil by Phillip Zimbardo

Darkness Visible: A Memoir of Madness by William Styron

Prozac Nation by Elizabeth Wurtzel

An Unquiet Mind: A Memoir of Moods and Madness by Kay Redfield Jamison

Awakenings by Oliver Sacks

The Man Who Mistook His Wife For a Hat: And Other Clinical Tales by Oliver Sacks

Thinking in Pictures: And Other Reports from My Life with Autism by Temple Grandin

Astonishing Hypothesis: The Scientific Search for the Soul by Francis Crick

Consciousness Explained by Daniel C. Dennett